Einstein Was Right

Einstein Was Right
The Science and History of Gravitational Waves

EDITED BY

Jed Z. Buchwald

PRINCETON UNIVERSITY PRESS
Princeton and Oxford

Published by Princeton University Press

41 William Street, Princeton, New Jersey 08540
6 Oxford Street, Woodstock, Oxfordshire OX20 1TR
press.princeton.edu

ISBN 978-0-691-19454-7
ISBN (e-book) 978-0-691-21197-8

British Library Cataloging-in-Publication Data is available

Editorial: Jessica Yao and Arthur Werneck
Production Editorial: Brigitte Pelner
Text and Jacket Design: Chris Ferrante
Production: Jacquie Poirier
Publicity: Matthew Taylor (US) and Katie Lewis (UK)

This book has been composed in Source Serif Pro and Futura PT

Printed on acid-free paper ∞

Printed in the United States of America

10 9 8 7 6 5 4 3 2 1

Contents

Preface

DIANA KORMOS BUCHWALD

The GR100 conference in Pasadena, California celebrating the centenary of the general theory of relativity and Albert Einstein's first paper on gravitational waves was many years in the making. The hope that gravitational waves would actually be directly detected by the year 2016 had been on our minds at least since the early 2000s, when the Einstein Papers Project, in collaboration with colleagues across the Institute, envisaged convening two large events separated by more than a decade that would bring together not only historians of science but also physicists, astronomers, and engineers working in relativistic fields to commemorate Einstein's 1905 *annus mirabilis* as well as his 1915–1916 papers on general relativity. But the post–9/11 restrictions on travel that prevented foreign scholars from easily attending US conferences meant that such larger gatherings would be international only in name. We therefore designed a series of individual lectures for 2004–2005 and proceeded to eagerly await news of gravitational waves over the next decade.

In April 2014, Kip Thorne, by then officially retired from Caltech, wrote Barry Barish:

> Dear Barry
>
> You may recall that a few months ago we talked about the possibility of an Einstein/general relativity celebratory event at Caltech. Diana Kormos-Buchwald has raised this issue with me again; see the attached email. If something were done in the 2016 time frame, it might coincide with the discovery of gravitational waves, or with putting improved limits on them. 1916 was Einstein's classic paper on gravitational waves. Thus, the natural connection to LIGO.
>
> Thanks and best wishes,
> Kip

That same day, Barish replied:

Kip / Diana,

Yes, I recall our earlier discussion and am enthusiastic about having this Einstein event at Caltech. I would be happy to step into your big shoes, Kip, and work with Diana to organize physics participation in the program, organization, etc.

If you agree, Diana and I can then arrange to get together, and then I will follow-up with Tom Soifer, Dave Reitze and others.

Barry

I visited Barish shortly afterward in his small office in West Bridge. Quiet chatter among those close to the inner circles of LIGO cognoscenti was by then indicating that good news was expected in the near future. But he was neither willing nor unwilling to declare by *when* gravitational waves would definitively be confirmed, though he seemed to suggest that the last week of Caltech's winter term 2015–2016, meaning mid-March 2016, would be a good bet. Barish accordingly held preliminary meetings with Caltech's Physics, Mathematics, and Astronomy Division (PMA) leadership. In September 2014, we followed up with Hirosi Ooguri, vice-chair of the Division. By October 2014, Jürgen Renn, expert in the history of general relativity, was selected by the Division of the Humanities and Social Sciences (HSS) as the recipient of the biannual Bacon Award in the History of Science.

From October 2014 to March 2016, the main organizers and a large number of colleagues exchanged more than 1,400 emails to bring the event celebrating the centenary of Einstein's gravitational-waves paper to fruition. Well-attended lectures were held both at the Huntington and at Caltech over a period of three days, despite occasionally heavy rain. *General Relativity at One Hundred: The Sixth Biennial Bacon Conference, 10–12 March 2016, Pasadena, CA,* was co-sponsored by the Bacon Foundation, the Walter Burke Institute for Theoretical Physics and the Divisions of the Humanities and Social Sciences and Physics, Mathematics and Astronomy at Caltech, and by the Research Department of the Huntington.

An exhibition documenting how Einstein formulated his theory of general relativity was on view during the month preceding the GR100 conference. The opening lecture on February 16 was delivered by Professor Hanoch Gutfreund, former president and chair of Theoretical Physics at the Hebrew University. The exhibition comprised facsimiles

FIGURE 0.1. Poster for the GR100 Conference, March 10–12, 2016.

General Relativity at 100: The Road to Relativity

Albert Einstein Manuscripts

On View in Dabney Lounge

Tuesday, February 16, to Monday, February 22, 2016, 9am-5pm

Opening Lecture: Dabney Lounge, February 16, 2016, 4pm

The Genesis of General Relativity: Documented and Visualized

Prof. Hanoch Gutfreund, former President & Andre Aisenstadt Chair of Theoretical Physics, The Hebrew University of Jerusalem

In celebration of the 100th anniversary of the theory of general relativity, the Division of the Humanities and Social Sciences and the Einstein Papers Project at Caltech present an exhibition of Einstein manuscripts, calculations, letters, books, and photographs housed at the Albert Einstein Archives at the Hebrew University in Jerusalem. The exhibit will contain manuscript highlights from the years 1907 to 1916, during which Einstein formulated general relativity and exchanged letters with many other scientists in Europe and the U. S.

The exhibit runs February 16-22 in Dabney Lounge after which time it will be moved to the Einstein Papers Project at 363 S. Hill Ave. on the eastern edge of the Caltech campus.

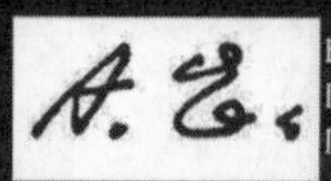

Einstein
Papers
Project

FIGURE 0.2. Poster of manuscripts and books exhibition preceding conference.

of selected highlights of Einstein's manuscripts from the years 1907 to 1916, calculations, letters to colleagues, photographs, and books, all materials housed at the Hebrew University of Jerusalem.

The conference also commemorated Albert Einstein's connections to Caltech, where he spent three academic terms as a visiting scientist during the years 1931 to 1933, the only US university at which Einstein held an appointment (albeit a temporary one). His links to the Institute reached back to 1913, when he famously inquired of George Ellery Hale whether the bending of light in the vicinity of the Sun could be observed during daytime.[1] Hale responded that the solar eclipse method was the only feasible one and, moreover, "In a short time, as soon as some additional data are available, I wish to ask your opinion regarding the theory of the general solar magnetic field which I have recently detected by observation of the Zeeman effect."[2] This early exchange concerning the Sun's, and the Earth's, magnetic fields would also preoccupy Einstein for several decades and constituted one of the main topics of discussion when he visited Pasadena almost twenty years later. Indeed, current work on Einstein's research during the 1920s has revealed that he hoped an explanation of the Earth's and the Sun's magnetic fields would provide insight into the possibility of formulating a unified theory of gravitation and electromagnetism. These aspects of his work are still poorly understood and have emerged only through recent publication of letters and manuscripts in the *Collected Papers of Albert Einstein*.[3]

1

Introduction

TILMAN SAUER

The first direct observational detection of gravitational waves by the Laser Interferometer Gravitational-Wave Observatory (LIGO) consortium on September 14, 2015, was a momentous event in the history of science. It shares a number of features with another decisive event of the previous century, the first direct observational detection of gravitational light bending during the solar eclipse of May 25, 1919. This, too, was a tiny effect predicted by a bold theory but was barely detectable given the day's technology. Its observational confirmation, like that of LIGO, took many years of preparation and involved failed attempts, its case made more difficult by the adversities of war and international hostility. Yet both investigations provided strong support for Einstein's theory of general relativity, a theory grounded on a willingness to question and seriously modify deeply entrenched notions of space and time.

Soon after the historic event took place, a meeting of leading scientists, historians, and philosophers was held at the California Institute of Technology to reflect on Einstein's legacy and to discuss its enduring validity. The meeting, which had been planned before the LIGO observation took place, reflected a centenary of Einstein's general theory. As it turned out, the detection occurred almost exactly a hundred years after Einstein published his foundational field equations. The meeting at Caltech, and the present volume, bring together leaders of the LIGO project with historians and sociologists of science to reflect on the event and its implications. The accounts in this volume offer a virtual participation in the process of science in the making, accompanied by informed historical, sociological, and philosophical reflection.

Barry Barish was principal investigator and director of the LIGO laboratory in its crucial period between the mid-1990s and 2005, when the project transitioned from a local endeavor at the two founding institutions Caltech and MIT to a multinational, multi-institutional large-scale science project involving eighteen nations, more than one hundred institutions, and in excess of a thousand individuals. His contribution opens the volume to remind us of the rapid pace of science, so rapid indeed that the detection overtook the initial planning of the anniversary conference at Caltech. Barish, one of the recipients of the 2017 Nobel Prize awarded for the discovery of gravitational waves, provides a concise account of the LIGO collaboration. His contribution conveys the sense of a rapidly expanding field that exploded into a spectrum of activities after a long period of preparation and gestation. Further observations of wave events followed the initial one on September 14, 2015. His words evoke a field in which one sensational, new event is hardly processed and prepared for public announcement when the next renders the first old news. Barish's contribution itself reflects the rapidly increasing success: he added a note in proof to announce that a second gravitational event (GW151226) had been detected during the first run (O1) of advanced LIGO, which took data between September 12, 2015, and January 19, 2016. That was followed by several more detections during the second observation run (O2) from November 30, 2016, to August 25, 2017. During that period, LIGO not only produced further evidence for binary black hole mergers but also yielded evidence for the merger of a binary neutron star as a first observation of a gravitational-wave event in conjunction with its electromagnetic counterpart. More new data can be expected from the third observation run (O3), which is projected to begin taking data in February 2019. This run will include, in addition to the two LIGO interferometers, data taken from the European observatory VIRGO, which should allow more accurate localization as well as give, for the first time, information about the polarization of gravitational radiation.

Kip Thorne, who with Barish and Rainer ("Rai") Weiss was awarded the 2017 Nobel Prize, was for many years a principal mover of the project directly to detect gravitational radiation, having begun theoretical investigations into the subject in the late 1960s. His contribution to this volume provides a broad perspective on the significance of the endeavor, emphasizing its importance for our understanding of the observable universe's curved space-time. Thorne's reflections beautifully convey his long-standing fascination with general relativity's counter-intuitive implications as he describes the development of the LIGO project from its first ideas to the large-scale international discovery machine. No one can tell this story better and with more authority than Kip Thorne, the institutional father of LIGO. Indeed, he has gone further in conveying the wonder of gravitational physics, serving as scientific adviser to the film *Interstellar,* which trades on the possibilities suggested by some of the field's implications.

Alessandra Buonanno, director at the Albert Einstein Institute for Gravitational Physics in Berlin and Professor of Physics at the University of Maryland, provides insight concerning what may begin to be detected with new gravitational-wave "telescopes." A principal aim of the extraordinary technology developed for the LIGO interferometer was and remains to provide qualitatively new information in the fields of astrophysics and cosmology. Buonanno sketches the possibilities now opened by gravitational-wave astronomy. She explains the intricate and fascinating astrophysical processes that take place when black holes or neutron stars collide, inspiral, merge, and settle down, and explains what gravitational and electromagnetic signals we can expect from those violent processes.

Dan Kennefick, professor of physics at the University of Arkansas and a longtime collaborator of the Einstein Papers Project at Caltech, provides an intimate, historical account of the LIGO project. In addition to training as a historian, while a graduate student Kennefick was early involved in the efforts led by Kip Thorne to prepare the theoretical grounds underlying the empirical search for gravitational waves that eventuated in LIGO's success. Drawing on his own experience and direct involvement, Kennefick points to the role of theorists and theoretical controversies in shaping the successful outcome of the quest, emphasizing in particular the importance of detailed numerical modeling. The large-scale simulations of such astrophysical catastrophes as black hole collisions through the explicit numerical solution of Einstein's equations provided the filters by which the raw interferometric data was interpreted. In 2005, a breakthrough became possible when it was realized that special coordinate conditions allowed the uninterrupted simulation of the full cycle of the inspiral, merger, and ring down of a binary black hole. That breakthrough allowed LIGO researchers to interpret the interferometric signals in terms of specific astrophysical causes, including estimates of distance as well as of initial and final masses.

Jürgen Renn, director at the Max Planck Institute for the History of Science, lays out a long-term history of research into the theory of relativity and gravitation that culminated in LIGO's successful detection of waves. He reaches back to the preconditions that underpinned the theory of general relativity and the details of the heuristics deployed by Einstein in his search for the field equations. That almost exactly a century passed between Einstein's publication of the equations and the first direct observation of ripples in space-time is, of course, a numerical contingency. But it is a contingency that appeals to the historical mind. Renn takes it as a challenge for the historian to account for the longue durée of a historical process that began with a theoretical conjecture, that underwent ups and downs in the structure's appeal and deployment by the physics community, settling at long last into a multimillion-dollar, large-scale endeavor that lasts for decades until delivering what had until then been a long-sought and discussed possibility. Renn focuses on the interplay between theoretical premises and experimental design, describing the transformation of the field from its origin in the imagination of a single mind to a collaborative enterprise involving thousands of scientists.

Harry Collins, professor of sociology at the University of Cardiff, has been interested in, and indeed associated with, the LIGO project for decades as a sociologist. Large-scale scientific projects, involving hundreds or thousands of researchers, technicians, and other personnel and enjoying levels of funding that surpass the means of individual groups or institutions, represent, Collins points out, a social reality of their own. This is particularly the case with LIGO, which was funded by the National Science Foundation at an unusually high level for more than two decades before achieving success. LIGO's efforts put ever more stringent constraints on the observability of gravitational-wave events as its instruments became steadily and impressively more accurate. Collins has observed the ongoing research with the eye of a critical and skeptical sociologist since the early seventies. He tells an intriguing story concerning an attempt by a LIGO predecessor, Joe Weber, to detect the effect. Given the technology available in the 1970s, Weber employed resonant bars. These large aluminum cylinders, Weber argued, would be set into vibration by passing gravitational waves. Despite the near unanimous rejection by the community of his early claims of detection, Weber, Collins argues, should be seen as a pioneer of the field because his experimental work created the community interest that made further work possible—most immediately the indirect observation of waves in 1975 by Hulse and Taylor due to energy loss by a neutron star binary.

Diana Buchwald, professor of history at Caltech and director of the Einstein Papers Project, focuses on Einstein's relationship with the California Institute of Technology. Not only is Caltech a founding institute and core partner of the LIGO endeavor, since 2000 it has been host to the long-term, multi-volume editorial project of the *Collected Papers of Albert Einstein*. Under Buchwald's aegis, to date eight volumes of writings and correspondence from the years 1918 to 1927 have been published. This is particularly apposite since Caltech's relationship with the founder of general relativity and the originator of the idea of gravitational waves goes back to the very beginnings of the theory, when Einstein corresponded in 1913 with the Caltech astronomer George Hale about the possibility of observing gravitational light deflection. Had historical circumstances been slightly different, Caltech might have become Einstein's home after he was forced to leave Germany and emigrate to the United States following the acquisition of power by the Nazis. He spent three winters in the early 1930s at Caltech, before finally accepting an offer at the Institute for Advanced Study in Princeton, which became his home for more than twenty years until his death in 1955.

Don Howard, professor of philosophy at the University of Notre Dame, focuses on the impact that Einstein's theory had for our modern understanding and philosophy of science. Philosophical reflection of science in the twentieth century underwent profound transformations with the advent of general relativity. Indeed, philosophy of science in a modern sense was created, Howard suggests, in its present form not the least by Einstein himself in a debate with philosophical interlocutors such as Moritz Schlick, Hans Reichenbach, Rudolf Carnap, Ernst Cassirer, and others. Howard goes back to the early years of the radically new understanding of space and time when it was first explored by philosophers. Howard describes how, in response, they reconsidered long-standing problems in the relationship between empirical and conceptual content, discussing in so doing the principal aspects of theory verification, the distinction between the a priori and convention, and the ontological structure of physical theory. Philosophy of science in the 1920s, Howard argues, reacted primarily and importantly to Einstein's theory of general relativity. In light of LIGO's first direct observation of gravitational waves, predicted by Einstein one hundred years before, this debate has lost nothing of its relevance.

2

The Quest for (and Discovery of) Gravitational Waves

BARRY C. BARISH

At Caltech, we began to discuss the program for this celebration of the hundredth anniversary of Einstein's Theory of General Relativity more than a year ago. At that time, we were just completing a major technical upgrade of LIGO to what we call Advanced LIGO. In fact, we had built and installed the Advanced LIGO components and were busy commissioning them for our first observational run. As is now public knowledge, we observed our first gravitational-wave event last September 11, days after we began data taking, and we announced the discovery and published the result in *Physical Review Letters* exactly one month ago (February 11, 2016).

My title for this talk was announced months ago, so I could not reveal the discovery, even to the organizing committee. But, today, at last I can alter my title from "The Quest for Gravitational Waves" to one that actually indicates the discovery! In this talk, I briefly describe LIGO, the improvements to Advanced LIGO that led to the discovery. I will finally discuss briefly what we discovered and a few implications.

I will begin when the National Science Board approved almost $300 million for the construction of LIGO in 1994. The initial version of LIGO was constructed during the period from 1994 to 2000 and employed technologies that represented a balance between being able to achieve sensitivity levels where the detections of gravitational waves might be "possible," while using techniques that we had demonstrated in our laboratories. LIGO was a huge extrapolation from the 30m prototype interferometer in Garching, Germany, and the 40m prototype interferometers at Caltech that preceded it, and especially considering the very large NSF investment, we needed to be confident of technical success. In reality, from the best theoretical estimates at the time, we anticipated that we would likely need to achieve sensitivities well beyond those of Initial LIGO before achieving detections.

This led us to a two-step concept for LIGO, the first being Initial LIGO, that used, as much as possible, proven technologies, while the second stage would significantly improve the interferometer sensitivities, using technologies that we needed to develop through an ambitious R&D program in our laboratories. It was from this perspective that we proposed to the National Science Foundation that, while we would be commissioning, running, and learning from Initial LIGO, we be funded to carry out an ambitious R&D program to develop techniques that would improve LIGO to a sensitivity where detection would become "probable." The NSF approved that plan and funded the successful R&D program beginning in about 2000 that led to a proposal for Advanced LIGO, which was submitted and the concept approved by the NSF in 2003. The actual project funding was approved several years later.

I would like to emphasize that the Initial LIGO infrastructure was designed such that the interferometer subsystems could be evolved or replaced inside the same infrastructure (vacuum vessels). After the completion and commissioning of Initial LIGO, we achieved much better sensitivity than previous gravitational-wave detectors and we took our first search data run. We did not detect gravitational waves but set new astronomical limits on various possible sources. Following the first data run, we made some technical improvements to Initial LIGO that reduced the background noise levels, and then we had a second data run. Again, we did not detect gravitational waves. We repeated this basic cycle for more than a decade, improving sensitivity and taking data runs, for a total of six data runs at ever-increasing sensitivity. For the final data runs, the interferometer sensitivities reached our original Initial LIGO design goals.

Unfortunately, even with this impressive interferometer performance we did not detect gravitational waves. The resulting limits on various sources of gravitational waves were more constraining for the models of gravitational-wave production in astrophysical phenomena, and we were cautiously confident that the improvements envisioned for Advanced LIGO would be sufficient to achieve detection.

At this point, the improved technologies developed for Advanced LIGO were mature enough to build and install in LIGO. The NSF funded the Advanced LIGO project once again through its Major Research Equipment and Construction (MREFC) program, and we undertook a multiyear major rebuild of LIGO to Advanced LIGO. The basic goal of Advanced LIGO was to improve the sensitivity from Initial LIGO by at least a factor of ten over the entire frequency range of the interferometer. It is important to note that a factor of ×10 improvement in sensitivity increases the distance we can search by that factor and therefore increases the volume of the universe (or rate for most sources) searched for by a factor of ×1000. (The sensitivity to most sources is proportional to the volume we search, therefore there is a very high premium in LIGO on increasing the range we can search, and we spend a good fraction of our time improving the sensitivity, rather than taking very long data runs.)

The planning for this meeting was fairly well advanced when construction was completed and commissioning of Advanced LIGO had begun. After several months, we improved the sensitivity beyond Initial LIGO by a factor of ×3 (×27 in rate) or more over most of our frequency band. This large increase in rate for gravitational-waves sources was certainly to motivate us to begin our first Advanced LIGO data run and defer further improvements until later. We embarked on a relatively short, three- to four-month run, that would already exceed all the running with Initial LIGO. We had just begun the run in September when, almost immediately, we discovered the Black Hole Binary merger event that we announced on February 11, 2016.[1]

Figure 2.1 is the discovery figure and is what we observed minutes after the event was detected in Advanced LIGO. The three figures each show the detected "strain" signals in units of 10^{-21} vs. time. The top trace is the observed waveform detected in the Hanford, Washington, interferometer, the middle trace is the observed waveform in Livingston, Louisiana. The two signals are almost identical but shifted by 6.9 msec and superposed in the bottom trace. The traces are almost raw data, only frequencies out of the bandwidth data have been removed. These waveforms have the characteristic form of a binary inspiral merger (as I will describe), and consequently this was identified as a candidate event by our LIGO collaborators at the Max Planck Institute in Germany within minutes of the actual passage of the gravitational waves. Of course, there was a great deal of excitement within the LIGO collaboration, but especially with a "new" detector, we had much work to do to establish whether this was a real signal. So, an intense period of analysis began almost immediately within the LIGO Scientific Collaboration.

Some History

Before describing this event further, I will briefly highlight some of the one-hundred-year history before this detection. Of course, the story began in 1915 when Einstein published his new theory of gravity, re-

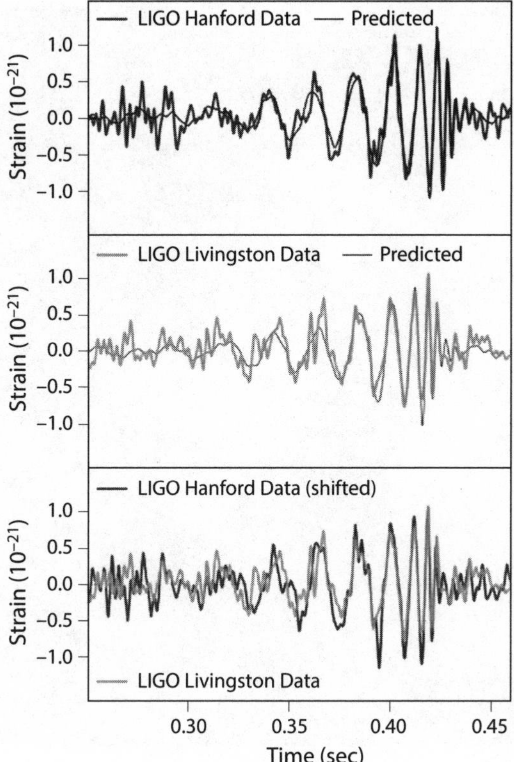

FIGURE 2.1. The discovery figure. Courtesy of Caltech/MIT/LIGO Lab.

placing Newton's theory, and the theme of this one hundredth anniversary conference. In Einstein's theory, the effects of gravity are through distortions of space-time, rather than Newton's simple description that the gravitational force is proportional to the product of the masses and inversely as the distance squared.

Newton's universal theory of gravity was enormously successful, describing a huge range of phenomena from the apple falling from a tree to the orbits of the planets. The immediate success of Einstein's theory of general relativity was that it solved the only known problem with Newton's theory, a ~10% discrepancy in the period of Mercury around the sun. This discrepancy had been a problem for about fifty years. The most popular view was that a still undiscovered object (or objects) existed between Mercury and the Sun and that that was responsible for perturbing the orbit of Mercury. No such object had been found in Einstein's time, or later, when NASA did an extensive search.

FIGURE 2.2. Albert Einstein. Courtesy of ÖNB / F. Schmutzer.

Einstein's new theory had the success of correctly predicting the period of the orbit of Mercury. However, fixing this one and only known problem with Newton's universal theory of gravity hardly warranted accepting a revolutionary new theory of gravity.

It is generally imperative that a new theory solve existing problems with an older theory, but also predict something new. This was the case for Einstein's theory of general relativity, by making a new prediction involving the bending of light as it passes near a massive object. That prediction was dramatically confirmed by Sir Arthur Edington in 1919.

The third prediction of Einstein's theory of general relativity was made the next year. In 1916, he predicted the existence of gravitational waves. This first paper on gravitational waves had an error of a factor of 2, which he fixed in a follow-up paper in 1918. It is important to note that, in the second paper, he introduced the quadrupole source term for gravitational waves.

Nevertheless, doubts persisted about whether gravitational waves really existed due to coordinate singularities that plagued theoretical formulations. Even Einstein had second thoughts. In 1936, after Einstein had moved to the United States, he and Nathan Rosen tried to develop the theory more quantitatively but found troubling singularities. As a result, they thought that gravitational waves were mathematical artifacts and did not exist. They submitted a paper to *Physical Review*, entitled "Do Gravitational Waves Exist?" Following a referee report pointing out their error, they resubmitted to the *Franklin Journal* and published a new formulation of gravitational waves in cylindrical coordinates. Similar doubts persisted in the theory community until the 1950s when the theoretical relativity community became convinced of their existence, especially at a conference in Chapel Hill, North Carolina, in 1958.

At this point, the problem of gravitational waves became an experimental rather than a theoretical question. Joe Weber, who was at the Chapel Hill meeting, was the first to take up the challenge of trying to detect gravitational waves. He developed a technique using resonant aluminum bars. Sensitive PZT detectors were mounted around the middle of the bar in order to detect distortions caused by the passage of gravitational waves. Each end of the bar acts like test masses in our interferometers, while the center acts like a spring.

The resonant bar technique was refined over many years. In fact, LIGO has inherited important legacies from Weber's work. The first is that he performed calculations of the limiting noise sources to guide how sensitive the detector could be and to focus the laboratory work to reduce noise, much like we do for LIGO. The second is that Weber employed two detectors placed far enough apart to make them independent of noise sources on the Earth, and this greatly reduced backgrounds by requiring a coincidence between the detectors within the speed of light. Finally, Weber used a scheme to determine the significance of observed signal candidates by comparing off-coincidence times to measure false coincidence background levels. All three of these techniques are important legacies we use in LIGO.

Although Weber's technical work and innovations led the way experimentally, he also claimed the discovery of gravitational waves several times, but none of these were confirmed in other detectors. The sensitivity of Weber's resonant bars was several orders of magnitude less than for the Advanced LIGO black hole binary merger that I report today.

Resonant bars continued to be improved and were formed into a worldwide network, only to be replaced by the large interferometers

FIGURE 2.3. The Livingston interferometer. Courtesy of Caltech/MIT/LIGO Lab.

when they became operational near the beginning of this century. The main reasons that resonant bars cannot achieve comparable sensitivities to interferometers is that they have a much narrower sensitive frequency band (near the resonant frequency) and because they have limited length of a few meters, compared to 4,000 meters in LIGO. (The signal strength goes like $\Delta L/L$.) Interestingly, one of Weber's students, Robert Forward, was the first person make a test interferometer for gravitational-wave detection.

The basic scheme for an interferometric gravitational-wave detector is to use a special high-power single-line laser beam (NdYAG) that enters an interferometer and is split into two beams that are transported in perpendicular directions. The vacuum pipe is 1.2m diameter and is kept at high vacuum. The "test" masses are mirrors that are suspended to keep them isolated from the Earth. They are made of fused silica and are hung in a four-stage pendulum for Advanced LIGO. The equal-length arms are adjusted such that the reflected light from mirrors at the far ends arrive back at the same time. Inverting one, the two beams cancel each other and no light is recorded in a photodetector. This is the normal state of the interferometer within how well background noise sources are controlled.

When a gravitational wave crosses the interferometer, it stretches one arm and compresses the other, causing the light from the two arms to return at slightly different times, and the two beams no longer completely cancel. This process reverses itself, stretching the other arm and squeezing the initial arm, back and forth at the frequency of the gravitational wave. The resulting light is recorded in a photo-sensor as the waveforms from the passage of a gravitational wave. The experimental challenge is to make the interferometer sensitive enough to the incredibly tiny distortions of space-time that come from a gravitational wave, while suppressing the various background noise sources.

The space-time distortions from the passage of an astrophysical source are expected to be of the order of $h = \Delta L/L \sim 10^{-21}$, a difference in length of a small fraction of the size of a proton. In LIGO, we have made the length of the interferometer arms as long as is practical, in our case four kilometers, and this results in a difference in length that is still incredibly small, about 10^{-18} meters. For reference, that is about 1,000 times smaller than the size of a proton! If that sounds very hard, it is!! Skipping the details, what enables us to achieve this precision is the sophisticated instrumentation that reduces seismic and thermal noise sources, by effectively making the statistics very high by having many photons traverse the interferometer arms.

LIGO consists of two identical interferometers, one in Livingston, Louisiana, and one in Hanford, Washington, about 3,000 km apart. When a gravitational wave crosses the Earth traveling at the speed of light, the time the signal is recorded in each interferometer will be within

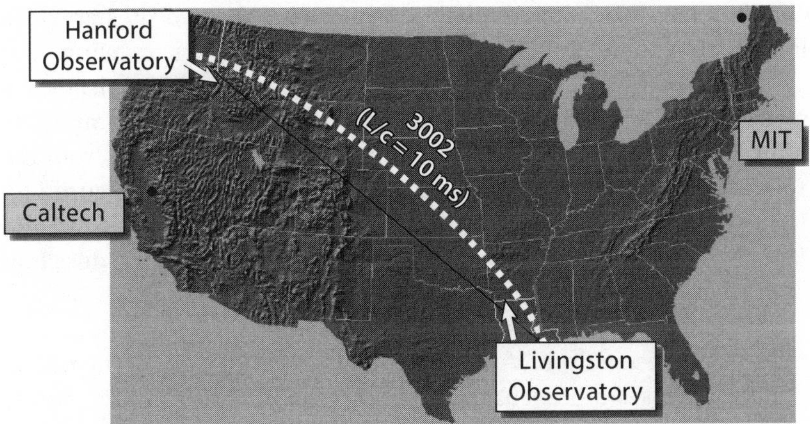

FIGURE 2.4. Locations of the Hanford and Livingston Observatories.

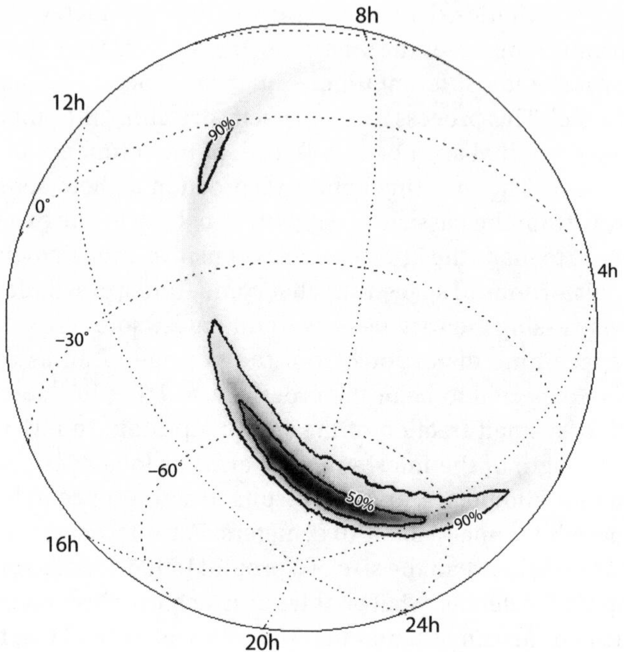

FIGURE 2.5. Southern hemisphere location of the signal source.

±10 msec, the maximum difference for waves traveling between the two detectors. We require a coincidence between the two interferometers within this time interval, and this greatly reduces backgrounds. In addition, the difference in the time of arrival at the two sites provides information on the direction the gravitational wave was traveling.

In the case of our observed event, the signal arrived in Louisiana 6.9 msec before arriving in Hanford, and combined with the amplitude information, we have located the source as having come up from the Southern Hemisphere within an area of 700 square degrees (figure 2.5). In the future, adding the Virgo detector in Italy, the KAGRA detector in Japan, and a LIGO detector in India will enable locating the direction of gravitational wave signals to far greater accuracy.

Figure 2.6 reveals the key features of a compact binary merger, as reflected in the analysis of the observed event. At the top of the figure, the three phases of the coalescence (inspiral, merger, and ringdown) are indicated above the waveforms. As the objects inspiral together, more and more gravitational waves are emitted and the frequency and

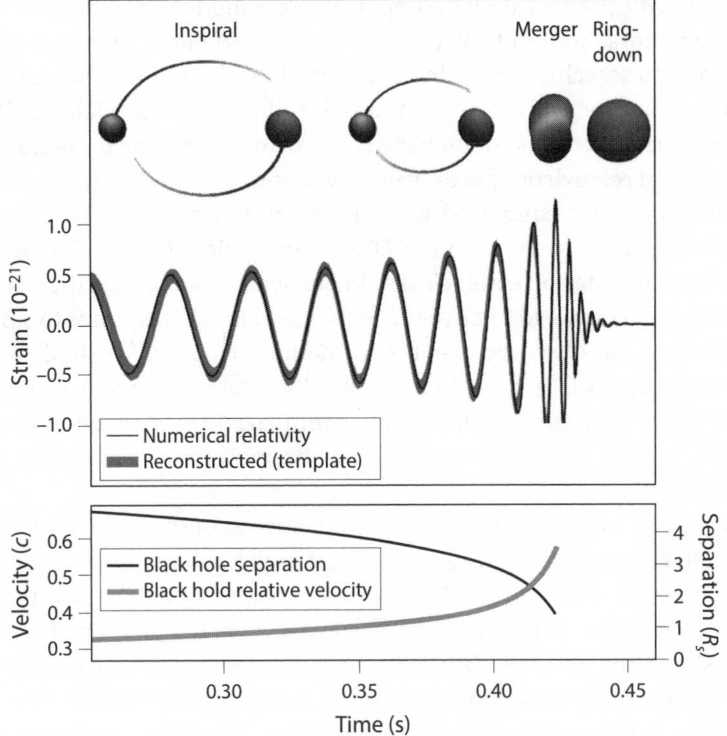

FIGURE 2.6. Features of a compact binary merger.

amplitude of the signal increase (the characteristic chirp signal), and following the coalescence, the merged single object rings down. The bottom pane shows (the left scale) that the objects are highly relativistic and are moving at more than 0.5 the speed of light by the time of the final merger. On the right side, the scale is units of Schwartzschild radii and indicate that the objects are very compact, only a few hundred kilometers apart when they enter our frequency band.

We conclude that we have observed two heavy compact objects, each approximately thirty times the mass of the Sun, going around each other at relativistic velocities and separated by only a few hundred kilometers. They merge due to the gravitational radiation coming from the accelerations. We conclude that this event is a merging system of two stellar black holes.

Finally, in order to be confident that we have observed a real event and not some sort of a background fluctuation, we directly measure the background probability by comparing coincidence time slices for the two detectors, both in time (e.g., ±10 msec) and out of time! That is, in addition to searching for coincidence in-time signals, we look for coincidences between all the out-of-time slices during our data taking. These background slices could not have come from any physical phenomena traveling at relativistic speeds, like gravitational waves. The total number of time slices we compared was equivalent to an in-time background level equivalent to more than 67,000 years of data taking. Taking into account the different event classes in our search, we reduce the limit on the false alarm rate to 1 in over 22,500 years. This corresponds to a probability that our observed event is accidental to $< 2 \times 10^{-6}$, establishing a significance level of 4.6 σ. I emphasize that the measured significance level is set from the number of bins compared from sixteen live days of data taking. This represents a lower limit on the actual significance of GW150911.

Figure 2.7 shows the statistical significance as described above for the GW150911 event, compared to the measured background levels under two different assumptions. The horizontal axis is a measure of the significance of the events and the vertical access is the rate. The left-hand plot shows the observed GW150911 event at the level of one event level, and having statistical significance σ > 4.6, as described above. This plot assumes a generic signal shape for the event. The right-hand plot shows a significance of > 5σ, when a binary coalescence form is used. Note that the second most significant event is about 2σ, which may well also be a binary black hole merger, but at this early stage in LIGO, we are only declaring five σ events as gravitational-wave binary mergers.

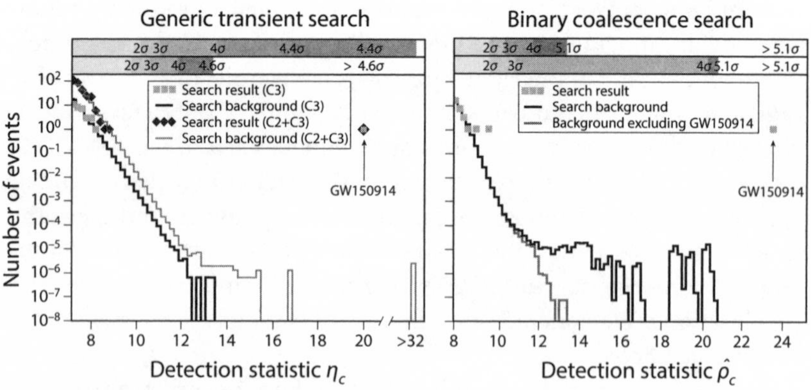

FIGURE 2.7. Coincidence slices for the two detectors.

The shape of the waveforms that describe the merger, coalescence, and ringdown reveal the detailed characteristics of the merger. The orbits decay as the two black holes accelerate around each other and emit energy into gravitational waves determined by the "chirp mass," as defined below.

$$M = \frac{(m_1 m_2)^{3/5}}{M^{1/5}} \simeq \frac{c^3}{G}\left[\frac{5}{96}\pi^{-8/3} f^{-11/3} \dot{f}\right]^{3/5}$$

The next orders allow for the measurement of the mass ratios and spins; the red-shifted masses (1 = z) m are directly measured; and the amplitude is inversely proportional to the luminosity distance. Orbital precession occurs when spins are misaligned with the orbital angular momentum. GW150911 shows no evidence for precession. The sky location, distance, and binary orientation information are extracted from the time delay between detectors and the differences in the amplitude and phase in the detectors.

Using numerical simulations to fit for the black hole merger parameters, we determine that the total energy radiated into gravitational waves is $3.0 \pm 0.5\ M_o\ c^2$. The system reached a peak energy of $\sim 3.6 \times 10^{56}$ ergs, and the spin of the final black hole < 0.7 of the maximal black hole spin. The main parameters of the black hole merger are summarized in table 2.1.

Table 2.1.

Primary black hole mass	$36^{+5}_{-4}\ M_\odot$
Secondary black hole mass	$29^{+4}_{-4}\ M_\odot$
Final black hole mass	$62^{+4}_{-4}\ M_\odot$
Final black hole spin	$0.67^{+0.05}_{-0.07}$
Luminosity distance	410^{+160}_{-180} Mpc
Source redshift, z	$0.09^{+0.03}_{-0.04}$

Last, with only two detectors we cannot locate the direction very well, but comparing the time, amplitude, and phase in the Livingston and Hanford interferometers, we are able to locate the gravitational wave as coming up from the Southern Hemisphere within an area of about 600 square degrees, as shown in figure 2.5. In the future, as we add the

Virgo, KAGRA, and LIGO-India detectors, we anticipate locating the source locations more than an order of magnitude better. In the long term, we expect that comparing gravitational-wave signals with electromagnetic signals, and maybe even neutrinos, will give rich information on the detailed dynamics and astrophysics. For black hole binaries, we don't expect electromagnetic counterparts, but we are hoping to detect black hole / neutron star and binary neutron star mergers in the not too distant future.

The dramatic discovery I have reported is the result of decades of detailed experimental work to develop instruments with the sensitivity capable of making these observations. The work on the interferometer facilities was by the LIGO Laboratory, managed in collaboration by Caltech and MIT, and including contributions from many scientists outside the laboratory. The science of LIGO is carried out through a worldwide collaboration, the LIGO Scientific Collaboration, of about 1,000 scientists.

Last, I note that the data run continued until January 12, 2016. During the same O1 four-month data run, but subsequent to this conference, we have reported the observation of a second event, GW151226.[2] This event was also from the merger of two black holes, but somewhat lighter than the event reported here. The different characteristics are beginning to give us information on the characteristics and populations of these binary black hole systems. We look forward as the detectors improve and allow us to get good statistics on black hole mergers which will help us to understand their origin and characteristics.

We eagerly anticipate a deeper understanding of our universe and, likely, many exciting surprises, as we open and exploit this new field of gravitational-wave science.

3

One Hundred Years of Relativity

From the Big Bang to Black Holes and Gravitational Waves

KIP S. THORNE

Newton, Einstein, and Their Frameworks for the Laws of Physics

In 1687, Isaac Newton gave us a framework for all the laws of physics that govern the universe, a framework that lasted for 218 years. It was based on the concepts of absolute space and absolute time, and on forces, accelerations, and other things of everyday experience.

In 1905, Albert Einstein gave us a *new* framework for the laws of physics, one that's now been in place for 111 years. Einstein called his framework the *Principle of Relativity*. It says that all the laws of physics must be the same in every freely moving laboratory everywhere in the universe. So his framework is actually a law that governs the laws of physics. That was audacious!

Einstein's framework has some amazing, counter-intuitive consequences. For example, if I measure the speed of light and get 300,000 kilometers per second, and you move past me at a speed of 200,000 kilometers a second, and you measure the speed of light, do you get the difference, 300,000 minus 200,000, that is, 100,000 kilometers per second? Our ordinary intuition about how speeds operate would say yes, you should measure 100,000. But Einstein's framework says NO, you must get the same as I got: 300,000.

How can that be? Something weird must be going on in space and time. And indeed it is. By exploring his framework deeply, Einstein concluded that you and I, moving at different speeds, must disagree about lengths, about times, and even about the concept of what events are simultaneous. So Einstein shook up the whole foundation of things that we thought we understood.

And then he looked at Newton's law of gravity, which says that the gravitational force by which the Sun pulls on the Earth varies inversely as the square of the distance between them. Einstein asked, the distance as measured by whom?

The Sun and the Earth disagree on their separation, because they're moving relative to each other. Hence, we have a quandary. Which distance should appear in Newton's law? That measured by the Sun, or the one measured by the Earth? This muddle shows, Einstein reasoned, that Newton's law of gravity violates my principle of relativity. Therefore, Newton must be wrong.

That was also audacious, since Newton then was universally recognized as the greatest scientific mind of all time. Here is Einstein, a very young man, not yet widely recognized as a genius, coming out and saying to the world that Newton was just plain wrong. Gravity has to behave differently.

Einstein's Search for a Relativistic Description of Gravity: The Warping of Time and Space; General Relativity

Einstein's next challenge was to find a whole new way to describe gravity, a way compatible with his principle of relativity. Jürgen Renn, in his beautiful Bacon lecture (which is included in this book), describes in careful detail how Einstein struggled to find his new laws of gravity, and the path that he followed. I'm going to give you an extremely simple version of this, one that I have chosen for fairly quick pedagogical clarity rather than faithfulness to Einstein's actual path.

Let me begin with what I like to call *Einstein's Law of Time Warps* (though that is not what Einstein called it): "Things like to live where they age the most slowly, and gravity pulls them there." (Isn't that where you would like to live?)

Einstein gave a beautiful mathematical formulation of this law, but I'll forego his math here. From his mathematical formulation, one can conclude that the Earth's mass warps time and this time warp produces the Earth's gravity. That's how Earth's gravity comes about. More specifically, time must slow by one second in 100 years on the Earth's surface compared to far from Earth, as that is the amount of slowing required to produce the gravity that we measure.

Now, that's not very much slowing. We don't gain an awful lot of extra life by living on the surface of the Earth. But that is the right amount of time slowing to produce the gravitational pull that we experience. In

1976, this quantitative prediction was verified to a precision of one part in 10,000, that is, 0.01%, when Robert Vessot of the Harvard-Smithsonian Center for Astrophysics flew an atomic clock in a rocket up to high altitude and compared its ticking rate to that of clocks back down on Earth.

Near a black hole, such as Gargantua in the movie *Interstellar* (I'll use *Interstellar* to illustrate several of Einstein's relativity predictions), gravity is enormously stronger than near Earth, so the slowing of time is enormously greater. In *Interstellar*, there's a planet (called "Miller's planet") that orbits near the surface (the "horizon") of Gargantua. One hour on that planet is the same as seven years on Earth. This enormous slowing of time produces gravity near Gargantua that is enormously larger than on Earth. This is illustrated compellingly in the movie. Cooper, speaking with his daughter, tells her that in his quest to save the human race, he may travel near a black hole and while there may age far more slowly than she does on Earth—so much more slowly that when he returns, she might be the same age as he, her father.

And that is what happens. When he emerges from the planet near Gargantua, she has grown up and become a brilliant theoretical physicist, while he has aged hardly at all. And then he goes down near the black hole Gargantua again, and returns to Earth and meets her. She is now a very old woman, and he still has hardly aged at all.

In this film, which was viewed by a worldwide audience of roughly a hundred million people, Einstein's Law of Time Warps truly came to life. That's one of the things that Christopher Nolan (the movie's director) and I wanted to achieve: to convey vividly some of the beautiful ideas, real science ideas, that are in general relativity.

In 1912, Einstein realized that, if time is warped, then space must also be warped. This warping of space was verified with high precision, again in 1976, by Robert Reasenberg and Irwin Shapiro, also of the Harvard-Smithsonian Center for Astrophysics. They led an experiment to measure the round-trip travel time of radio signals that go from the Earth to the *Viking* spacecraft (which was then in orbit around Mars) and back to Earth. Mars carried the spacecraft near the Sun as seen from Earth and then away, so the rays along which successive radio signals traveled on their journey from Earth to spacecraft and back were as shown in the bottom left of figure 3.1. Early in the experiment the rays traveled far from the Sun; later, quite near; and still later, far again. Reasenberg and Shapiro discovered that the round-trip travel times along these rays were not what you would expect if the space between the Earth and the spacecraft had been flat. There was an extra time delay in the waves' return (upper left

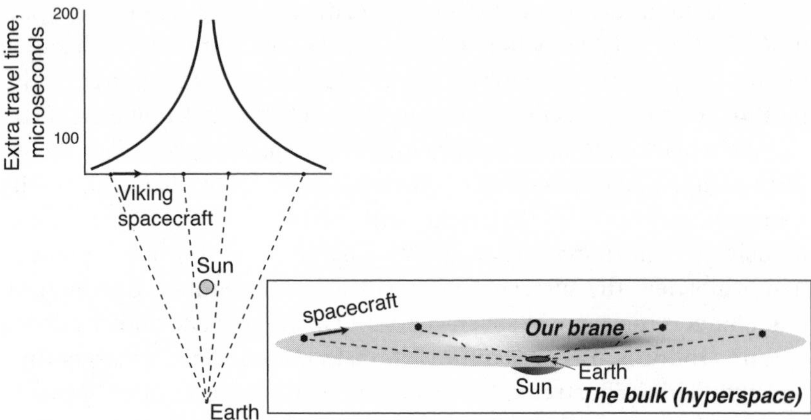

FIGURE 3.1. Reasenberg-Shapiro Experiment that measured the warping of space around the Sun using round-trip radio signals between Earth and the *Viking* spacecraft. Adapted from *The Science of Interstellar* by Kip Thorne.

of figure 3.1), which was small along rays that are far from the Sun, but grew as large as 200 microseconds when the waves passed near the Sun. From this and the fact that the speed of light (and radio signals) is always constant (according to Einstein's principle of relativity), they concluded that the *distances* that the signals traveled were longer than if space were flat. So space had to be warped. And they could infer the precise shape of the warped space from how the delay changed as the rays changed.

We can visualize this space warp by taking the plane formed by all the rays and embedding it in a three-dimensional flat space. Looking in from that flat space, we can see the warpage (lower right of figure 3.1). The warpage of space, inferred in this way, agreed beautifully with the predictions of Einstein's general relativity theory.

(As a side remark, in the movie *Interstellar*, our universe with its three space dimensions and one time dimension lives inside a "hyperspace" or "bulk" that has one more space dimension—the "fifth dimension" of the movie. If we ignore time, and focus on the surface formed by the radio rays of the Reasenberg-Shapiro experiment, then we get precisely the picture in the lower right of figure 3.1. *Thus, we can think of that picture as showing the warped space around the Sun as seen from the bulk.*)

Between 1912 and November 1915, Einstein struggled to discover the law that controls the warping of space and of time. On November 25, 1915, he finally figured it out. Today we call that law *Einstein's Field Equation*, and it has the deceptively simple form

$$G_{\mu\nu} = 8\pi G T_{\mu\nu}.$$

Amazingly, once you know the meanings of this law's mathematical symbols and the physics that they embody, you discover that it contains almost everything there is to know about Classical Physics (non-Quantum Physics).

Consequences of Einstein's General Relativity Theory

In the remainder of this chapter, I will describe a few of the many things Einstein's equation has taught us in the past century.

If I had all day, I would talk about a huge number of fascinating consequences, among them:

- high-precision experimental tests of general relativity, in the solar system and in binary pulsars
- cosmology: the big bang, inflation, the cosmic microwave background radiation, the origin of galaxies and clusters of galaxies, dark matter, dark energy
- geometrodynamics: the nonlinear dynamics of warped spacetime
- gravitational waves: the opening of a wonderful new window onto our universe
- the incompatibility of general relativity and quantum theory: the quest for new laws of quantum gravity—string theory, M theory, loop quantum gravity
- speculations: cosmic strings, wormholes, backward time travel, macroscopic higher dimensions as in *Interstellar*

But as most of these are beyond the limitations of this chapter, I have chosen out of all of these topics just a few things that I find especially wonderful. My idiosyncratic choices are all topics that involve spacetime warps that occur without the aid of any matter whatsoever.

Black Holes

A black hole is the quintessential example of this: It is an object made entirely from warped spacetime. Figure 3.2 is a picture of what a non-spinning black hole would look like as seen from the bulk. In other words, it is an equatorial slice through the black hole, as seen embedded in a flat three-dimensional space, the bulk. The horizon of the black hole (its surface, out of which nothing can ever escape) is the black circle at the bottom. When we switch from visualizing an equatorial slice to the full three-dimensional black hole, that circle becomes a sphere (as the physicist Romilly explains in *Interstellar*); the horizon is actually a sphere.

As seen from the bulk, the black hole's space looks like a trumpet horn, and like the surface of a whirlpool. Far from the horizon (far off the printed page), the space asymptotes to a flat sheet, the distant universe. In figure 3.2 (see color plate 5), the space is color coded to depict the slowing of time that controls the black hole's gravity. In the yellow region, time flows at 10% of the rate far away (on Earth); at the horizon, time slows to a halt, and that infinite slowing makes gravity at the horizon be infinitely strong, preventing anything from getting out of the hole.

What produces this warping of space and time? It's not matter. There's no matter in the black hole at all. No material stuff of any sort. It's true the black hole was created by the implosion of a star long ago. But the star's matter is long since gone, destroyed at the center of the black hole; it no longer exists. With the matter gone, the only thing that can produce warping is the energy of the warping itself! The black hole is a *gravitational soliton*: an object held together by the "nonlinear influence" of the energy of its own warpage.

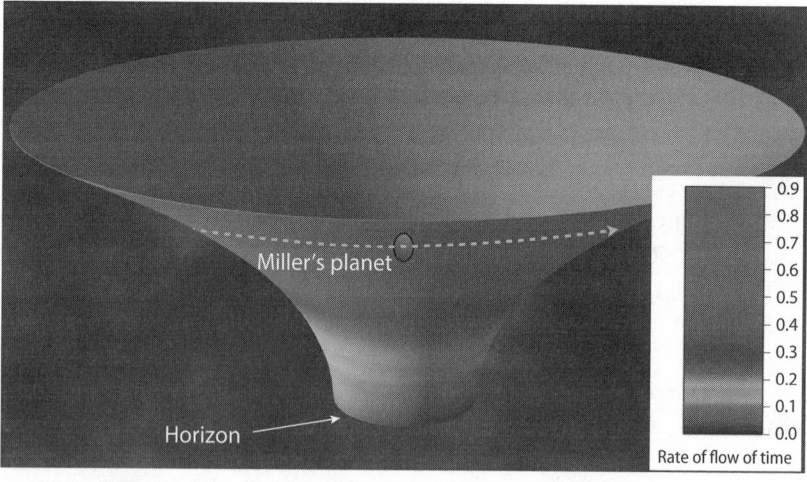

FIGURE 3.2. Equatorial slice through a non-spinning black hole, as viewed from the bulk, together with the smallest orbit that Miller's planet can have and not plunge into the hole. Adapted from a drawing by Don Davis, based on a sketch by Kip Thorne. Courtesy of NASA/JPL.

The water planet in *Interstellar*, "Miller's Planet," cannot be any closer to the black hole than the orbit shown in figure 3.2, because if it were closer, the orbit would be unstable and the planet would spiral into the black hole and be gone. That orbit is so far away from the horizon, that the slowing of time is quite small—by contrast with the huge slowing depicted in the movie. Consequently, it is impossible for the slowing depicted in *Interstellar* to be scientifically accurate if the black hole is nonspinning, as in figure 3.2.

But if the black hole spins very fast, then its spin drags space into a whirling motion around itself like the air in a tornado (figure 3.3, see color plate 6), and that whirl of space stabilizes the orbit of Miller's planet, so the planet can be down very close to the horizon and survive, as shown in figure 3.3. For one hour on the planet to be seven years on Earth, as in *Interstellar*, the spin must be far higher than seems reasonable astrophysically. But it is possible; it is not ruled out by the laws of physics, and Christopher Nolan wanted one Miller hour to be seven Earth years. So we gave the black hole Gargantua that very high spin.

In *Interstellar*, two gigantic water waves wash over Cooper's *Ranger* spacecraft on Miller's planet. Each wave is solitonic: It holds itself together by a nonlinear self-interaction, just like the black hole holds itself together by the nonlinear effects of its own energy of warping. The second wave arrives about an hour after the first.

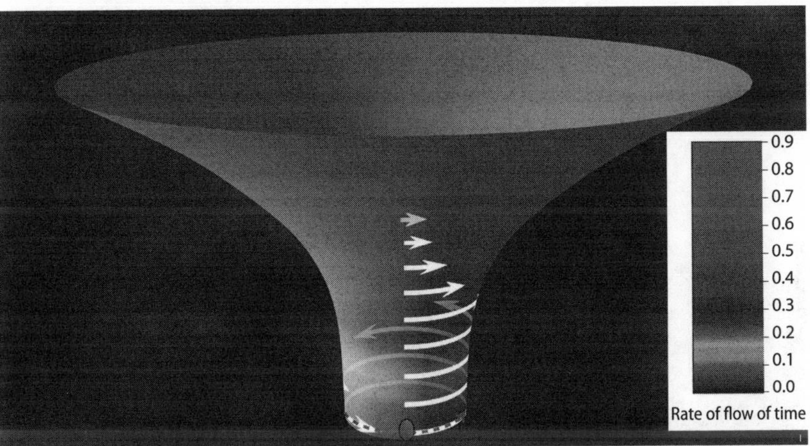

FIGURE 3.3. The warped space of the black hole Gargantua, which spins very rapidly. The white arrows depict the hole's dragging of space into a whirling motion, which stabilizes the orbit of Miller's planet, permitting it to be very close to the hole's horizon. Adapted from a drawing by Don Davis, based on a sketch by Kip Thorne. Courtesy of NASA/JPL.

What could possibly generate such water waves? The answer is the *tidal gravity* of the black hole Gargantua. This is the same type of gravity by which the Earth's moon creates the tides on Earth's oceans. In the left side of figure 3.4, the bottom of Miller's planet is closer to Gargantua's horizon than the top, so Gargantua pulls more strongly on the bottom than on the top, and as a result the planet gets stretched. And similarly, it turns out, the planet gets squeezed from the sides, and so deformed as shown. This same stretch and squeeze of the Earth's oceans (produced by the Moon's and Sun's gravity) results in the Earth's ocean tides. But for Miller's planet the stretch and squeeze are enormously stronger than for the Earth, since they are produced by a very nearby black hole.

In *Interstellar,* the planet has somehow been deposited into its near-horizon orbit quite recently, as seen by the planet, though long ago as seen from Earth, where time flows far, far faster. The planet is rocking back and forth, as shown in the right side of figure 3.4, gradually settling down toward an equilibrium state with the same face always pointed toward Gargantua. As the planet rocks, its oceans slosh, producing the giant waves seen in the movie.

Einstein's general relativity field equation governs the deformation of Miller's planet and also governs its rocking. The deformation and rocking are strongly influenced by Gargantua's mass. If Gargantua is much lighter than 100 million suns, the tidal forces will tear Miller's planet apart. In *Interstellar*, Christopher Nolan and I wanted the planet to be strongly deformed but not torn apart, so we chose the mass of Gargantua to be that of 100 million suns. From that mass, I compute that the time required for one rock, back and forth, of Miller's planet is about an hour, so this is the time between the giant water waves.

It's really quite wonderful, I think, that Chris was able to get all this relativity physics embedded into his movie in such a graphically com-

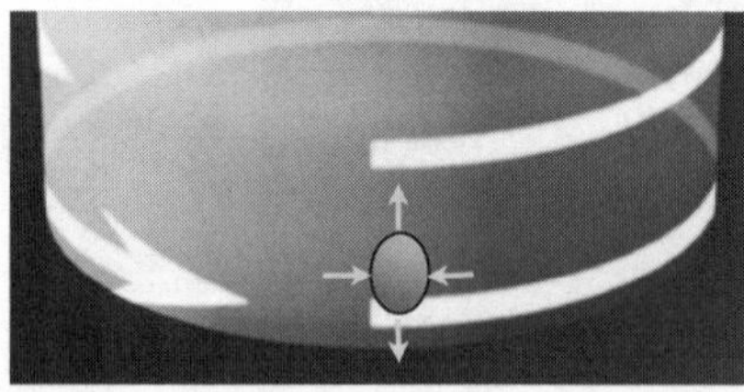

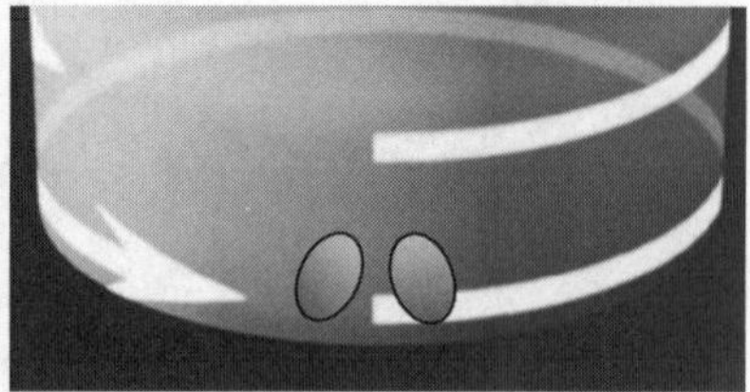

FIGURE 3.4. *Left.* Miller's planet is deformed by Gargantua's tidal gravity. *Right.* The deformed planet rocks back and forth, producing a sloshing of the planet's oceans that results in giant water waves. Adapted from a drawing by Don Davis, based on a sketch by Kip Thorne. Courtesy of NASA/JPL. See color plate 7.

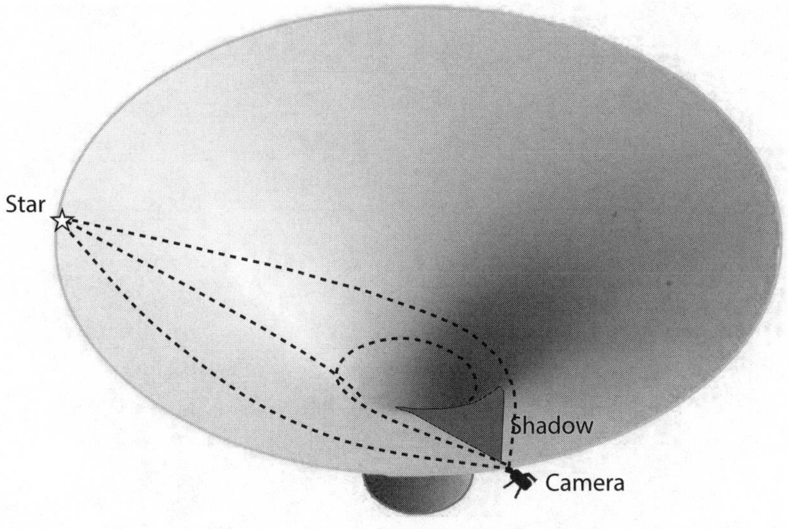

FIGURE 3.5. Three light rays that travel along three different paths though the warped space of a black hole, from a star to a camera. Adapted from an image in *The Science of Interstellar* by Kip Thorne. See color plate 8.

pelling way: the slowing of time, the sloshing of water due to extreme tidal gravity, and more, much more.

Particularly iconic in the movie is the appearance of a black hole as seen by human eyes (or an iMax camera). To understand this, begin with a camera looking at images of a star that is far from the hole, as depicted in figure 3.5. If spacetime were not warped, there would be just one light ray from the star to the camera: the straight line between them. But because of the warping, there are many such light rays; the figure depicts three of them. The camera sees an image of the star coming in along each light ray, so three images in the figure, but a huge number in reality. This is called gravitational lensing, and when there are many stars, and many images for each star, it gives rise to a remarkable pattern of swirling stellar images as the camera orbits around the black hole. A snapshot from a movie[1] of those swirling images is shown in figure 3.6. The pitch black, nearly circular region in the center is the shadow that the black hole's horizon makes in front of the field of stars that are being gravitationally lensed. The strange flattening and thin striations on the left side of the shadow are caused by the whirling motion of space, which is toward the camera on that left side.

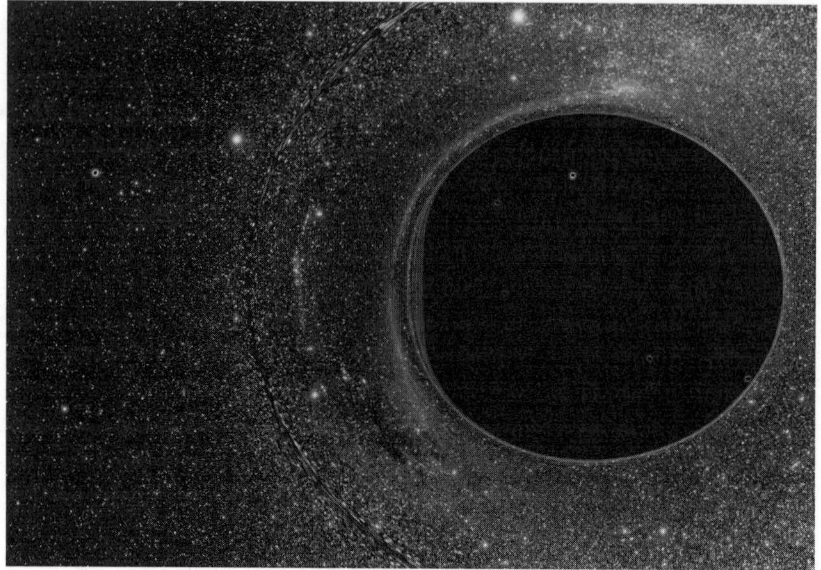

FIGURE 3.6. A field of many stars gravitationally lensed by a fast-spinning black hole. Courtesy DNEG. See color plate 9.

Accretion Disks around Black Holes

But figure 3.6 is not what we saw in *Interstellar*. What we saw was the black hole Gargantua surrounded by a disk of hot gas that gradually accretes onto the black hole, a disk so bright that it blinds us to the stars (figure 3.7). This has since become the iconic black hole image in popular culture.

How does this image come about? The disk of hot gas is thin and lies in Gargantua's equatorial plane (upper right of figure 3.8). The iMax camera is a bit above the disk's plane, as shown in the figure. Light rays from the upper back face of the disk, for example ray A, travel up over the black hole and down to the camera; they are pulled around the hole and down by the hole's intense gravity. These light rays produce the upper piece of the disk image (piece A in the lower left of the figure). Similarly, light rays from the disk's lower back face, for example ray B, travel under the black hole, which pulls them upward to the camera, producing the lower piece of the disk image (piece B in the lower left of the figure). And rays such as C, from the front of the disk, travel directly to the camera, producing the central bar in the image (piece C in the lower left of the figure). So it's all very simple: again, the culprit is

FIGURE 3.7. A variant of the image of the black hole Gargantua with a thin disk of hot gas in its equatorial plane, as seen in *Interstellar*. From James, Von Tunzelmann, Franklin, and Thorne (2015), © 2015 IOP Publishing Ltd. under CC BY 3.0 License. See color plate 10.

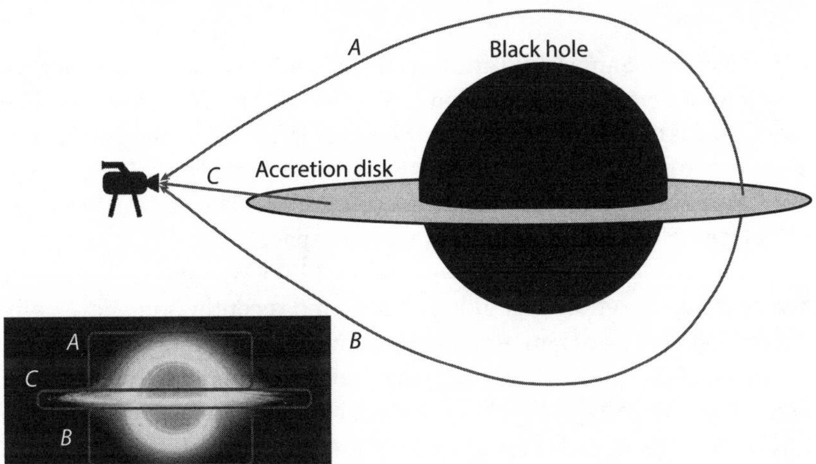

FIGURE 3.8. The iconic image of the black hole Gargantua is produced by gravitational lensing of light from its thin, equatorial accretion disk. Adapted from *The Science of Interstellar* by Kip Thorne, using the variant of Gargantua's image shown in figure 3.7.

gravitational lensing—a natural result of the warping of space and time, and of the gravity that the time warp produces.

Where do accretion disks like this one come from? Many are produced when the tidal gravity of black holes tears apart stars that stray too close. Figure 3.9 is an image from a movie of such a *tidal disruption*—a movie made by astrophysicist postdocs James Guillochon (University

FIGURE 3.9. Hot gas from tidal disruption of a star by a black hole. Courtesy of James Guillochon. See color plate 12.

of California at Santa Cruz) and Suvi Gezari (Johns Hopkins University), based on a computer simulation that solves Einstein's equations. The black hole is the tiny black spec in the upper left crook of the gas stream. The orange gas stream is the remnant of the disrupted star. The gas near the hole is beginning to form an accretion disk. The gas farther from the black hole is escaping into interstellar space.

There is a black hole in the center of our Milky Way galaxy. It has only a very weak accretion disk. It hasn't been fed recently. Someday, it will get fed and will have a nice rich disk for a while.

Andrea Ghez at UCLA and her team have been mapping the orbits of stars in the vicinity of this black hole for more than twenty years; see figure 3.10. The black hole is at the location of the star . . . the focus of all the approximately elliptical orbits. From the details of the orbits, Ghez and her group deduce that the black hole weighs four million times what the Sun weighs—a relatively light black hole compared to the one in Gargantua. By contrast, the black hole at the center of the nearest big galaxy to our own, the Andromeda galaxy, weighs 100 million times more than our Sun: the same as *Interstellar*'s Gargantua.

For the black hole at the center of our Milky Way galaxy, radio astronomers are likely to actually image the accretion disk and the black hole's shadow in the next several years. They will do so using a set of radio telescopes that are linked together to make a single near-Earth-sized telescope, called *the event horizon telescope*. This is a marvelous collab-

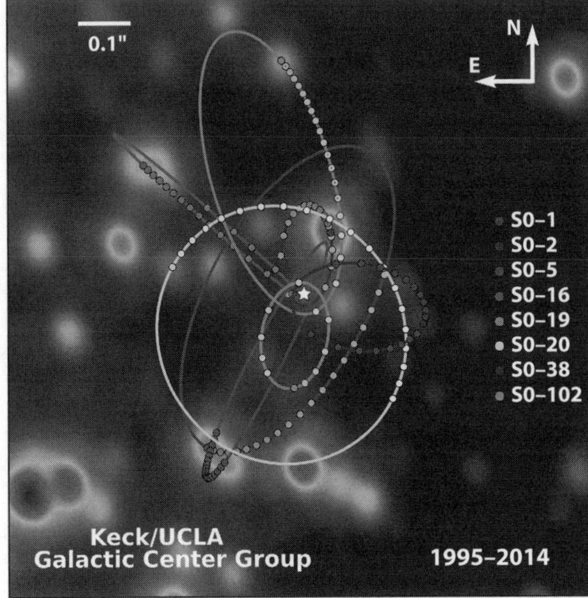

FIGURE 3.10. Orbits of stars around the black hole at the center of our Milky Way galaxy as mapped by Andrea Ghez. Courtesy of Keck / UCLA Galactic Center Research Group, Andrea Ghez. See color plate 13.

oration of hundreds of astronomers from thirty-four universities and institutes. Note added in proof: In 2019, the event horizon telescope successfully imaged a far larger black hole in the more distant galaxy M87.

Inside a Black Hole: The Flow of Time, Three Singularities, and the Laws of Quantum Gravity

What is inside a black hole? Let me approach this gradually. Suppose, first, that you are at the surface of a black hole, its event horizon, hovering there, not falling in (left spacecraft in figure 3.11). Then time will stop for you, whence gravity there is infinitely strong, so you can only hover momentarily; you are pulled inexorably into the hole and cannot escape.

As you plunge into the hole, you might expect the time you feel to be slower than stopped time. That, of course, is nonsensical. So, what really happens to your time? The answer is that inside the black hole, time flows in a direction you would have thought was a space direction: It flows downward, toward the hole's center and toward *singularities* that reside there. Once you cross the horizon, you are dragged downward by the forward flow of time (right spacecraft in figure 3.11). That's another explanation of why you can't get out of a black hole: neither you nor

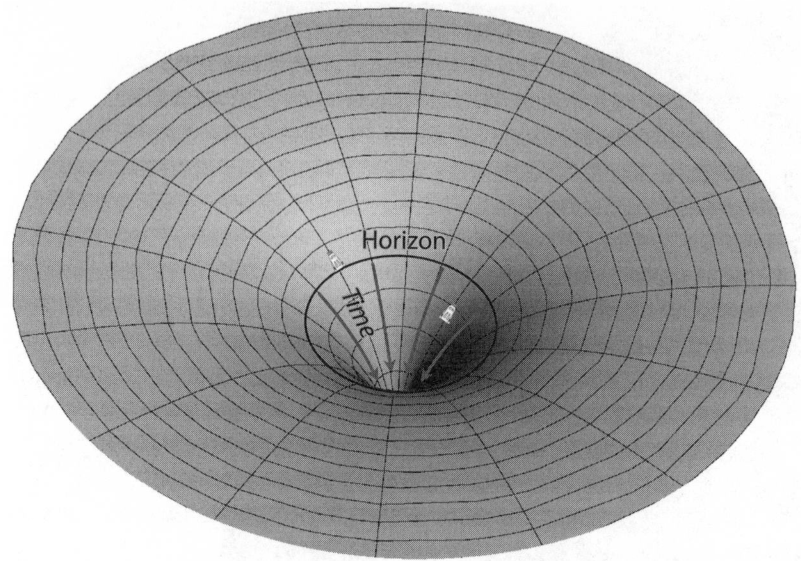

FIGURE 3.11. A black hole as seen from the bulk. Time slows to a halt in a spacecraft hovering at the horizon. Inside the black hole, time flows downward. Drawing by Kip Thorne.

anything else can travel against the forward-and-downward flow of time, according to Einstein's general relativity theory.

In *Interstellar*, Cooper, played by Matthew McConaughey, plunges into Gargantua. Nothing special happens to him as he crosses the horizon. He feels no infinite gravity, because he is falling. Tidal gravity does not suddenly become huge; it changes only slowly, gradually. When he looks upward, he sees the universe above, and the accretion disk. He can see out of the hole because light from above falls through the horizon and onto him. But people outside the hole cannot see him, since he cannot send signals upward against the flow of time.

Deep inside the black hole, according to general relativity, there is a chaotic singularity (left side of figure 3.12): a region where tidal gravity becomes infinitely strong in a chaotic way, stretching first in one direction then another, and then another, faster and faster, in a chaotic pattern. This is called the BKL singularity, so named for the three Russians who discovered it by solving Einstein's field equation: Belinsky, Khalatnikov, and Lifshitz. If Cooper hits the BKL singularity, he will be killed by the infinitely growing tidal forces, and then the atoms of which his body are made will be torn apart and destroyed.

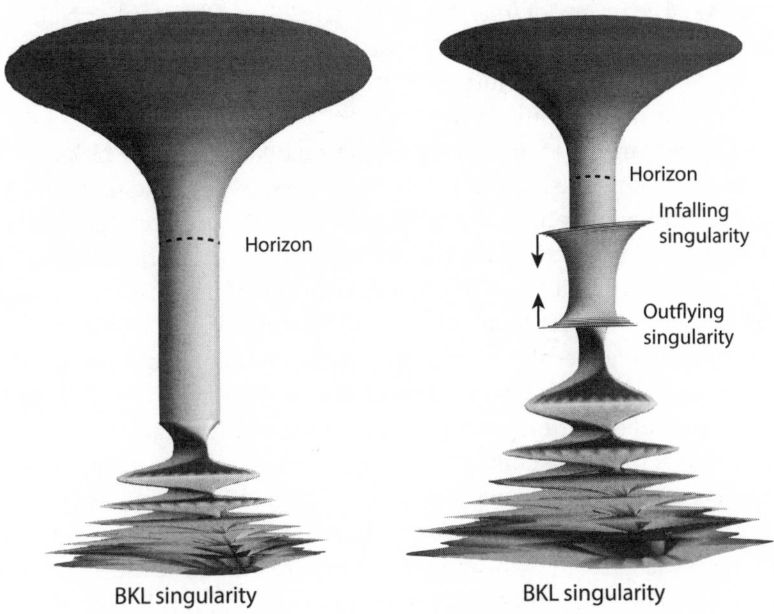

FIGURE 3.12. *Left:* Heuristic drawing of the chaotic BKL singularity inside a black hole, as seen from the bulk. *Right:* The infalling and outflying singularities produced by stuff that falls into the black hole. From *The Science of Interstellar* by Kip Thorne. See color plate 15.

Now, it turns out there are actually three singularities inside a black hole. In addition to the chaotic BKL singularity, there are two others that are more gentle (right side of figure 3.12). When Cooper falls into Gargantua, he gets caught between the other two. One is falling inward toward him. It is created by everything that will fall into the black hole in the future, over billions of years. Time is so screwed up inside the black hole, that all that infalling stuff gets compressed into a thin sheet, with infinite tidal gravity, that descends toward Cooper at near light speed. The other singularity is flying upward toward Cooper. It is created by small portions of the stuff that fell into the black hole in the past, portions that got scattered back upward. That outflying singularity also has tidal gravity that grows infinitely strong, though in a much more benign way than the BKL singularity.

In *Interstellar*, just before Cooper hits the outflying singularity, he gets scooped up by an alien spacecraft with four space dimensions, called the *tesseract*, and carried into the bulk; and he survives. If you want to understand the weird things that happen after that, you'll have to read my book, *The Science of Interstellar* (Thorne 2014). Christopher Nolan so admires Stanley Kubrick's film, *2001: A Space Odyssey*, that he chose to make his own movie's ending as inscrutable as Kubrick's—and invited me to explain the ending in my book.

Physicists are eager to understand the singularities inside black holes, because the singularities' ultimate behaviors, as their tidal gravity becomes infinitely strong, are controlled by the laws of quantum gravity. (Cooper's goal in plunging into Gargantua is to extract information about these quantum gravity laws through measurements that he and the robot Tars make near the outflying singularity.) These laws arise from some sort of synthesis of general relativity and quantum theory, and the quest to learn them has been the Holy Grail of theoretical physics since about 1960.

The laws of quantum gravity not only control the singularities inside black holes; they also controlled the birth of our universe. So, once we understand the laws of quantum gravity, we should be able to use them to understand how the universe was born, and whether there are other universes besides our own, and what if any are the connections between universes. The quantum gravity laws also control, I think, whether you can build a machine to take you backward in time. We physicists suspect that every time machine, no matter how it is designed, will self-destruct when it is activated; see, for example, Hawking et al. (2002). This is Stephen Hawking's *chronology protection conjecture*. Of course, building time machines and activating them is far beyond our technological capabilities today. But Hawking and I and others have struggled to figure out what general relativity predicts for the fates of time machines that are made by infinitely advanced civilizations. Hawking and I like to make bets with each other, but in this case we made no bet, because we came to agree that a time machine's fate is controlled by the laws of quantum gravity, and will remain inscrutable until physicists understand those laws with confidence.

Gravitational Waves

Gravitational waves are another prediction of general relativity. These waves, like black holes, are made wholly and solely from warped spacetime. I have devoted much of my career to gravitational waves, and so for the remainder of this chapter I shall discuss them from my own personal, historical perspective.[2] This book of written lectures contains three other chapters about gravitational waves, one by Barry Barish, whose leadership of LIGO was crucial to the discovery of these waves, another by Alessandra Buonanno, whose contributions to theory and data analysis were also crucial, and a third by the sociologist of science Harry Collins. My lecture and this chapter are complementary to theirs.

Gravitational waves can be thought of as ripples in the shape of space that travel at the same speed as light. They exert tidal-gravity forces on any object through which they pass, for example Miller's planet (figure 3.13). They have no influence along the direction in which they propagate (into the page in the figure), but they stretch the planet along one transverse direction and squeeze it along the other. The stretch and squeeze are oscillatory, and the pattern of oscillations—the waves "waveform" (figure 3.14)— contains information about the waves' source.

Gravitational waveforms are rather like those for sound. You can feed sound into an oscilloscope and watch, on a screen, the sound's oscillating waveform; and similarly, you can feed the output of a gravitational-wave detector into a computer to produce an image of the gravitational waveform. And just as the sound waveforms from a symphony carry rich information about the orchestra, what its various instruments are doing,

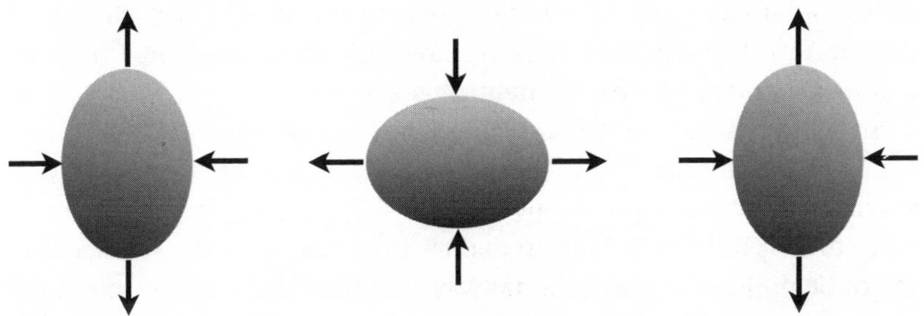

FIGURE 3.13. A gravitational wave propagating perpendicular to the plane of this picture oscillates between horizontal squeeze and vertical stretch, and horizontal stretch and vertical squeeze.

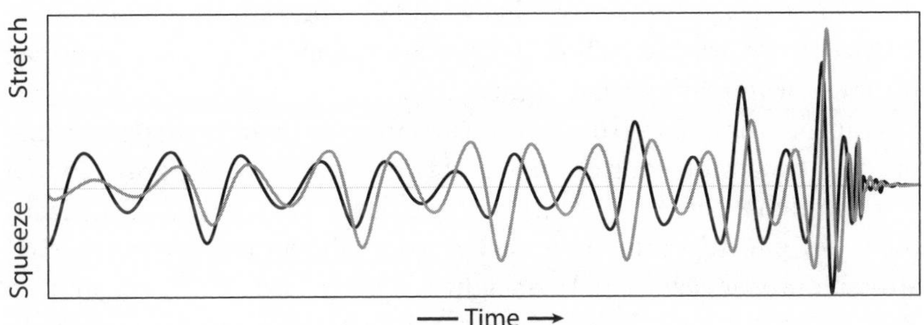

FIGURE 3.14. Two gravitational waveforms, one red and one black, produced by black holes that orbit each other and then collide. Courtesy SXS Collaboration.

and the emotions of the composer, so also the gravitational waveforms can carry rich information.

Gravitational waves were predicted and elucidated by Albert Einstein in 1916, using his general relativity theory. For the first half century after Einstein's prediction, theorists struggled to understand them (and also struggled to understand black holes). Are they real, physical phenomena, or are they figments of the mathematics that have no true physical reality? For some details of this struggle, see Barry Barish's lecture in this volume, and also see Daniel Kennefick's beautiful book, Kennefick (2007).

And during that first half century most physicists, including Einstein, thought the waves coming to Earth from the astrophysical universe would all be so weak that they might never be detected.

Joseph Weber, at the University of Maryland, in the late 1950s and 1960s was the first physicist to have the courage and inventiveness to design and build gravitational-wave detectors, and search for waves from the astrophysical universe, but he did not find any. See Barish's chapter for details. I met Weber in 1963 at a summer school in the French Alps, where I was a student and he lectured. Through his lectures and through conversations as we hiked together, I became enamored of gravitational waves. And through John Wheeler, my PhD thesis advisor, I became enamored of black holes and neutron stars.

So it was natural that when I joined the Caltech faculty in 1966, I created a research group that focused on black holes, neutron stars, and gravitational waves. Most important for this chapter, in 1972, with my PhD student Bill Press, I began to develop a vision for the science that might be done with gravitational waves, if they could be detected: the kinds of information that might be extracted from the waves, and what astrophysicists might do with that information.

In parallel, also in 1972, Rainer (Rai) Weiss at MIT was developing a design for a new type of gravitational-wave detector—a *gravitational-wave interferometer* of the sort that would ultimately be implemented in LIGO. In a remarkable technical paper dated April 15, 1972, Rai described his interferometer's design (figure 3.15).

Four mirrors hang from overhead supports in an L-shaped configuration: two at the ends of the L and two at the corner. A gravitational wave stretches one pair of mirrors apart ever so slightly and squeezes the other pair together, and a laser beam (shown red) measures that stretch and squeeze. The laser light is split in two at a beam splitter. Each resulting beam bounces back and forth between the mirrors of its arm. Then the beams recombine and interfere with each other at

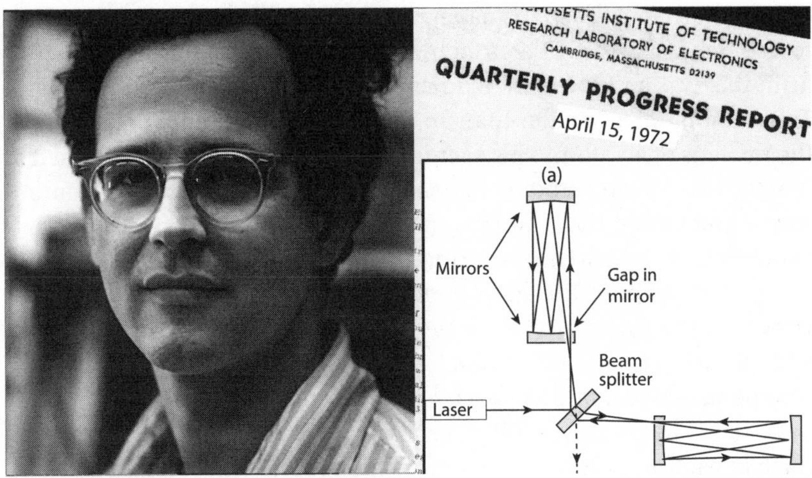

FIGURE 3.15. *Left:* Rai Weiss ca. 1972. *Lower right:* The bare bones of Rai's proposed gravitational interferometer, as seen from above. *Upper right:* The beginning of Rai's technical paper about this interferometer. Courtesy Rainer Weiss.

the beam splitter. The interfered light going downward in the picture enters a photodetector that measures its intensity. As the arms stretch and squeeze, that intensity rises and falls due to the changing difference in the travel distance of the interfering beams. The multiple bounces in each arm increase the difference in travel distance, thereby increasing the light intensity's rise and fall.

In Rai's technical paper, he identified all of the major noise sources that such an interferometer would have to face, explained ways to deal with each and every one of those noise sources, and estimated how big the resulting total noise would be. By comparing with the strengths of the waves that my astrophysicist colleagues and I were predicting, he concluded that, if the interferometer were a few kilometers in size, it would have a real possibility to detect the predicted waves. His analysis was an amazing tour de force.

Rai did not publish this technical paper in a standard scientific journal such as *Physical Review* or the *Astrophysical Journal*, where the rest of us were publishing. No. Rai, being Rai, thought he shouldn't publish until he had built his interferometer and shown it would work, and perhaps even detected gravitational waves! So instead, he put his paper in an internal MIT technical report series (upper right of figure 3.15), and then sent copies to many colleagues around the world, including me.

When I first perused his paper, I was highly skeptical. Rai proposed to use light to measure a stretch and squeeze of the interferometer's arms that was nearly a trillion times smaller than the light's wavelength! Ten million times smaller than an atom! a hundred times smaller than the nucleus of an atom! This seemed crazy to me—until I discussed it at length with Rai and studied his paper in depth. Then I became enthusiastic and vowed that I and my theory students would do whatever we could to help Rai and his experimenter colleagues succeed.

In 1976, at about the same time as I was making this vow, Ronald Drever at the University of Glasgow became enamored of Rai's ideas, and following Rai's lead, started building an interferometer. In the process he invented some important improvements on Rai's design. Most important, he proposed a different way to bounce the light back and forth in each arm—a way that is more compact and has turned out to be more versatile, but that is much more difficult to implement, technically (figure 3.16). In technical language, he made each arm into a Fabry Perot cavity. In other words, he made the round-trip distance in each arm be an integer multiple of the light's wavelength, so the light can bounce back and forth resonantly inside the arm. The light gets "sucked" into each arm, gets trapped there for many bounces, then leaks back out and interferes at the beam splitter.

In 1976, having become convinced that gravitational-wave detection was likely to succeed, I proposed that we create an experimental effort at Caltech to build and perfect these gravitational-wave interferometers, in parallel with Rai's effort at MIT, Drever's new effort at Glasgow, and an effort recently initiated near Munich, Germany by Heinz Billings. My proposal was embraced enthusiastically by Caltech's faculty, provost, and

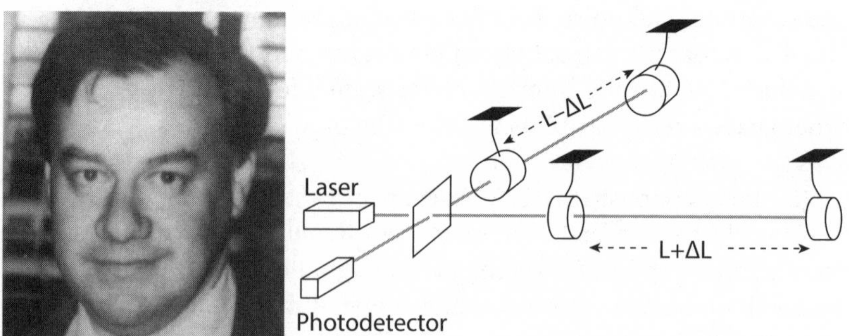

FIGURE 3.16. Ronald Drever (photo) and his version of Rai Weiss's gravitational interferometer. Courtesy Caltech Archives.

president; and I was even invited to talk to the Caltech trustees about it. The trustees didn't have to approve it, but they were enthusiastic. The entire Caltech community embraced our gravitational-wave quest and stuck with us enthusiastically right up to our ultimate success, forty years later.

In 1979–80 we brought Ronald (Ron) Drever to Caltech, from Glasgow, to lead our effort, and brought Stan Whitcomb from Chicago as an assistant professor to co-lead it. The Caltech administration provided approximately $2 million (in 1979 money!) to get the effort off the ground, and I think that played a large role in convincing the US National Science Foundation (NSF) to begin investing substantial sums. Under Ron's oversight, Stan led the Caltech experimental team in building a prototype interferometer with 40-meter-long arms (just 1% of the arm length that we would ultimately use in LIGO); figure 3.17. (Stan went on to become chief scientist in the LIGO Laboratory and one of the several most effective and influential LIGO experimenters in our 40-year quest.) In parallel, in 1980–83, the NSF funded Rai to continue work on his smaller prototype, and most important, to do a feasibility study for the kilometer-sized interferometers that we knew we would ultimately need.

In 1984, Rai, Ron, and I presented to a key NSF committee the results of Rai's feasibility study, and the achievements of the prototypes at Caltech, MIT, Glasgow, and Munich. Our progress was great enough that NSF encouraged us to initiate plans for 4-kilometer-long LIGO interferometers.

For three years, 1984–1987, Rai, Ron, and I led this planning effort. Ours was among the most dysfunctional leaderships that the physics community has ever seen, so in late 1986 our NSF program officer

FIGURE 3.17. Stan Whitcomb (middle) and colleagues working on the Caltech 40m prototype interferometer ca. 1982. Courtesy Caltech Archives.

(an amazing physicist named Richard Isaacson who had made major contributions to the theory of gravitational waves) told us that, in order to move forward we must get a strong, single leader: a single director for LIGO. And so, with the help of the Caltech and MIT administrations, we recruited Rochus (Robbie) Vogt to lead us.

Robbie had created Caltech's Space Radiation Laboratory to conduct cosmic ray research and had made a major personal mark on that field. He had been the first chief scientist at Caltech's Jet Propulsion Laboratory, and he had been Caltech's chair of Physics, Mathematics and Astronomy, and Caltech's provost—and he was tough. He "knocked heads together" in the Caltech and MIT LIGO teams, making them truly merge and work together effectively for the first time. And he led us in writing a superb construction proposal for LIGO. Our proposal contained a detailed vision and plan for first building LIGO's facilities—the vacuum system and so forth in which our interferometers would operate—and then building two generations of interferometers. The first interferometers would be at a sensitivity where we would have to be awfully lucky to see anything, but they would give the experimental team enough experience and insights to then build the far more complex second generation: advanced interferometers that are sensitive enough to likely see lots of gravitational waves.

Our proposal was reviewed by several hard-nosed NSF committees and got high marks, so NSF seemed ready to move forward. But then we ran into a buzz saw. LIGO was being funded out of the Physics Divi-

FIGURE 3.18. Me, Ron Drever, and Robbie Vogt ca. 1988. Courtesy Caltech Archives.

sion of NSF. It was being designed and built by physicists for whom our two-generation approach is quite natural. But our ultimate goal was to do astronomy. And there were a number of eminent members of the astronomy community who could not accept the idea that you build something costing nearly $300 million that's unlikely to see anything, and then you spend more money on second-generation instruments that you claim will have success. For $300 million, you could create a fabulous optical telescope, or several. As a result, we got blindsided in congressional testimony by an attack from eminent astronomers and were driven into an intense political battle.

NSF, Caltech, and MIT stood firmly by us through this battle, as did a significant portion of the astronomy community, and Robbie led us with wisdom. Finally, after two years of struggle to educate key members of Congress and their staffs, Congress bought in and provided our first major funding for LIGO. Congress has stood by us firmly from then (1992) through the detection of our first gravitational waves (2015)—for twenty-three years—and beyond, regardless of who was in power, Democrats or Republicans.

As we moved forward toward construction, NSF and Caltech got cold feet over the very lean management structure that Robbie was planning, and so Robbie stepped down as LIGO director and was replaced by Caltech's Barry Barish, who had extensive experience in leading large, high-energy physics projects. Barry, in fact, is widely viewed as the most effective leader of large projects that the physics community has ever seen, and LIGO's success is due in very large measure to his leadership. In his chapter in this volume he describes in detail the history of LIGO from 1994, when he became LIGO director, up to the present.

FIGURE 3.19. Barry Barish. Courtesy Caltech Archives.

In extreme brevity, beginning in 1997, Barry expanded LIGO from about fifty scientists at Caltech and MIT to about a thousand at about eighty institutions in fifteen nations. This expansion was critical to success, as the LIGO interferometers are so complex that building and operating them successfully requires far more skilled scientists and engineers than Caltech and MIT alone could provide. Barry led us in building the first generation of interferometers and carrying out their first searches (2000–2005), and then turned over LIGO's leadership to Jay Marx and then David Reitze, who steered us through subsequent gravitational-wave searches with the initial interferometers (2005–2010), and the construction and installation of the advanced interferometers (2010–2015).

In parallel with this major experimental effort, theorists were working hard to predict the gravitational waveforms that LIGO was likely to see, particularly the waveforms from colliding black holes, which I was pretty sure would be LIGO's strongest source. These waveforms would be long (many cycles of oscillation) and could be complex and difficult to find in LIGO's noisy data. To find them with confidence, we needed a large catalog of their possible wave shapes, which would underpin the analysis of LIGO's data. And to extract the information carried by the waves, we would need to compare the observed waveforms with the waveforms predicted for black holes with various masses and spins.

In her chapter in this volume, Alessandra Buonanno sketches how the required waveform catalog was, in the end, built and then used in the data analysis. The black holes in a binary gradually spiral together emitting oscillatory waves, then collide in a great cataclysm, producing a huge final burst of waves. During the gradual inspiral, the waveforms can be computed using so-called post-Newtonian techniques (largely with pencil and paper). But the only way to analyze the collision and its huge final burst of waves is by solving Einstein's field equation numerically, on a very large computer. This is called numerical relativity.

Theorists began working on numerical-relativity simulations of colliding black holes in the 1950s, at about the time that Joseph Weber was planning his first gravitational-wave detector. But progress was very slow. In the early 2000s, I became worried that the simulations would not produce the needed waveforms in time for LIGO's first wave detections. Fortunately, by then I had trained a number of young theorists who could take over the roles that I was playing in LIGO, so I left day-to-day involvement with the LIGO project to focus on building a numerical relativity effort at Caltech, as an adjunct to an already existing effort at Cornell University led by Saul Teukolsky. We called our effort the *SXS collaboration*, for *Simulating eXtreme Spacetimes*. Under Saul's leadership, it has now been expanded to several other universities, it includes roughly fifty computational physicists, and it has had great success. Saul is now a joint professor at Caltech and Cornell.

By September 2015, when LIGO captured its first gravitational waves, our SXS team had built a catalog of several hundred waveforms from black hole collisions—each waveform produced by a pair of black holes with a specific ratio of the holes' masses, and specific values of the holes' vectorial spins; see figure 3.20. (The magnitude of the stretch and squeeze and the duration of the waves are both proportional to the holes' total mass, so we don't have to specify the total mass in the simulations.) Our catalog was used to underpin the LIGO data analysis, as Alessandra Buonanno describes in her chapter.

This brings me to September 14, 2015. The LIGO team was preparing for its first gravitational-wave search with the advanced LIGO interferometers, when, unexpectedly, a strong burst of gravitational waves arrived—a burst that the LIGO team has named GW150914 after the date it arrived. In their chapters in this book, Barish and Buonanno describe that first gravitational-wave discovery from the viewpoint of LIGO scientists. I will describe it from the viewpoint of the waves themselves and their source—the source that the LIGO team identified with the aid of the SXS simulations.

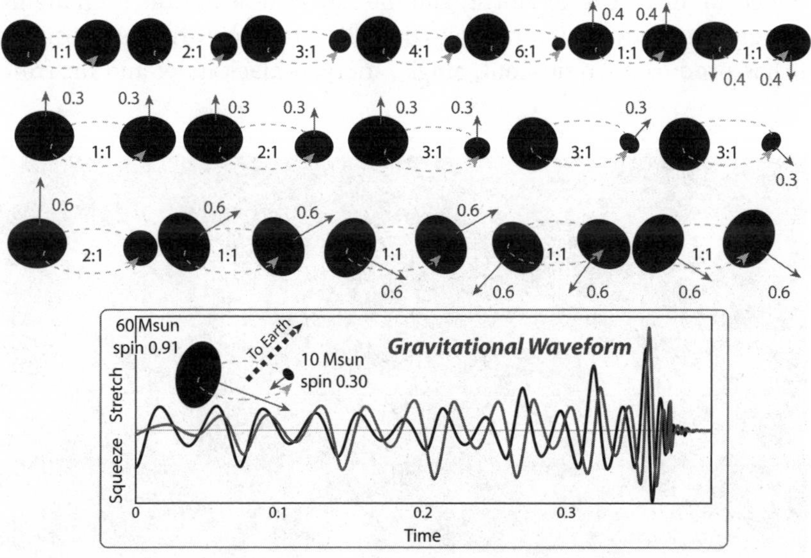

FIGURE 3.20. *Top:* Seventeen of the black hole binaries whose waveforms are in the SXS catalog; 4:1 means a mass ratio of 4 to 1, the red arrows indicate the directions and magnitudes of the black holes' spins, the red numbers are the spin magnitudes as a fraction of the maximum possible spin, and the green arrows depict the orbit. *Bottom:* The two waveforms for a specific example of the binary's parameters. (A gravitational wave has two waveforms, just as light has two polarizations; LIGO measures only one of them.) Courtesy SXS Collaboration. See color plate 17.

Approximately 1.3 billion years ago, when here on Earth multi-celled life was just spreading around the globe, but in a galaxy far, far away, two black holes spiraled around and around each other, emitting gravitational waves. Figure 3.21, based on an SXS simulation, shows what those black holes would have looked like to you, if you had been nearby. This is an analog of figure 3.6: it shows the shadows of the two black holes' horizons in front of a field of many stars and shows the swirling pattern of stellar images produced by the holes' gravitationally lensing the stars' light rays.

As they lost energy to gravitational waves, the two holes gradually spiraled inward and then collided, creating a veritable storm in the shapes of space and time. Figure 3.22 depicts that storm, as seen looking in on our universe from the bulk; this is an analog of figures 3.2 and 3.3. The top picture shows the binary black hole (BBH) 60 milliseconds before collision. The space around each black hole dips downward like the water surface in a whirlpool, and the color shifts from green to red (time slows) as one moves down the tube. The middle picture shows the BBH at the moment of collision. The collision has created a veritable *storm* in the shape of spacetime: Space is writhing like the surface of the ocean in a weather storm, and the rate of flow of time is changing rapidly. The bottom picture shows the BBH after the storm has subsided. It has produced a quiescent, single, merged black hole; and far from

FIGURE 3.21. The black hole binary that produced LIGO's first gravitational-wave signal, GW150914, as your eyes would have seen it if you had been nearby. Image produced by Andy Bohn, Francois Hébert, and Will Throwe from an SXS simulation. Courtesy SXS Collaboration. See color plate 18.

the hole, a burst of gravitational waves (depicted only heuristically as water-wave-type ripples) flows out into the universe.

The initial black holes weighed 36 and 29 times as much as the Sun, for a total of 65. The final black hole weighed only 62 times the Sun, so the collision converted three solar masses of black hole mass into gravitational waves. It was as though Nature had annihilated three suns and converted their masses entirely into gravitational-wave energy. And this was done so quickly, in less than a tenth of a second, that the total power output during the collision (the total energy emitted per unit time) was fifty times larger than the total power output of all the stars in the universe put together! It was the most powerful explosion that astronomers have ever seen.

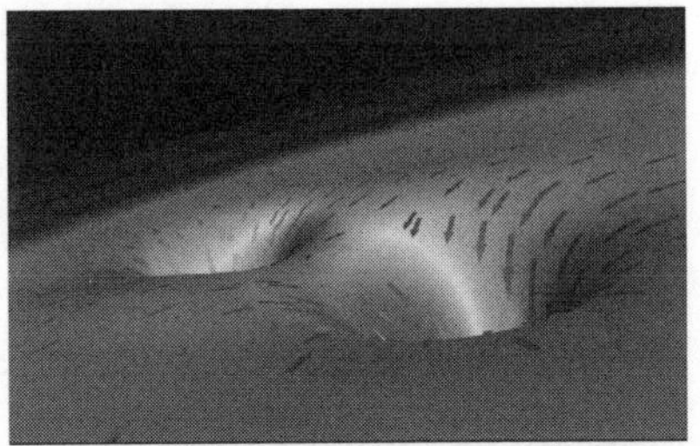

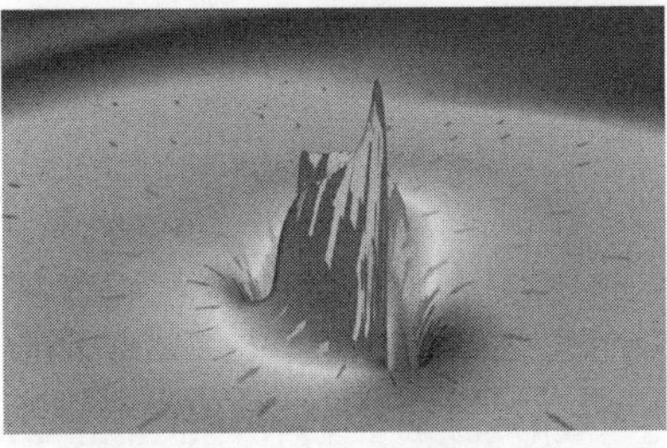

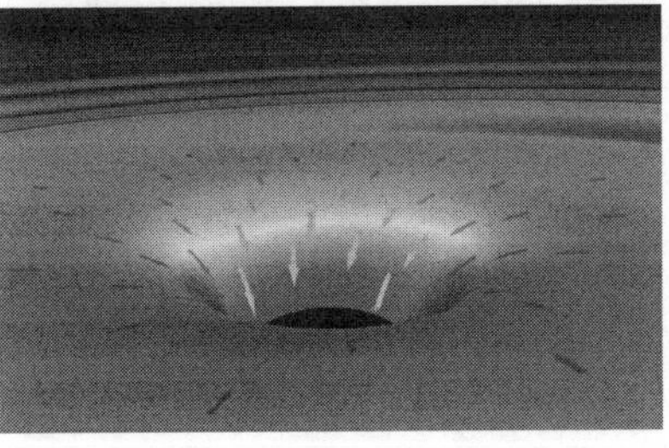

FIGURE 3.22. Snapshots from a movie depicting the geometry of spacetime around the GW150914 binary black hole, as seen from the bulk, 60 ms before collision, at the moment of collision, and 12 ms after the collision. Movie generated by Harald Pfeiffer from an SXS simulation. Courtesy SXS Collaboration. See color plate 19.

The gravitational waves traveled out of the galaxy in which the black holes lived, into the vast reaches of interstellar space. For 1.3 billion years they traveled, until 50,000 years ago, when our ancestors were sharing our Earth with the Neanderthals, the waves reached the outer reaches of our Milky Way galaxy. They traveled onward for 50,000 years, arriving at Earth on September 14, 2015. They arrived first near the tip of the Antarctic peninsula, then traveled upward through the Earth, unscathed, emerging at Livingston, Louisiana, where they stretched and squeezed one LIGO interferometer, and then 7 milliseconds later emerging at Hanford, Washington, where they stretched and squeezed the other LIGO interferometer.

Our LIGO team, assisted by members of the European Virgo gravitational-wave team, analyzed the signal for about four months to be absolutely sure it was truly produced by a gravitational wave and to deduce the details of its source, and then, on February 11, 2016, we announced our discovery: humans' first encounter with a gravitational wave. For far more detail, see the chapters by Barish and Buonanno in this book.

GW150914 is just the beginning. Over the next decade, as LIGO's sensitivity improves, it will see many hundreds of binary black holes, hundreds of colliding neutron stars, and perhaps hundreds of black holes tearing apart companion neutron stars—and also perhaps the core of a supernova explosion, and defects in the structure of space called cosmic strings. LIGO is searching for all of these. But most interesting of all will be huge surprises: Whenever a new method has been devised for observing the universe, surprises have come; and gravitational waves are so radically different from other ways of observing that huge surprises are almost guaranteed.

LIGO observes gravitational waves with periods of 1 to 100 milliseconds. Over the next two decades, three other types of gravitational-wave detectors operating in three other period ranges will reach maturity and begin probing the universe. Space-based gravitational-wave interferometers such as the European Space Agency's LISA mission will observe waves with periods of minutes to hours. Radio telescopes, by monitoring arrays of pulsars, will observe gravitational waves with periods of roughly three to thirty years. And by mapping the polarization patterns of cosmic microwaves over the sky, astronomers will indirectly observe gravitational waves from the earliest moments of our universe with periods today of hundreds of millions of years. By the middle of this century, I expect gravitational-wave astronomers to be exploring our universe's birth and, with the aid of their observations, physicists will be mastering the laws of quantum gravity that governed our universe's birth.

Four hundred years ago Galileo built a small optical telescope, pointed it at Jupiter, and discovered Jupiter's four largest moons—and thereby initiated instrument-based electromagnetic astronomy. In 2015, the LIGO team turned on its advanced interferometers and discovered gravitational waves from colliding black holes—and thereby initiated gravitational- wave astronomy. Over the four hundred years since Galileo, electromagnetic astronomy has totally revolutionized our understanding of the universe. I invite you to speculate about what gravitational-wave astronomy will bring over the next four hundred years.

4

The New Era of Gravitational-Wave Physics and Astrophysics

ALESSANDRA BUONANNO*

From the extreme dynamics of black holes to the beginning of the universe itself, the detection of gravitational waves in 2015 has opened a rich new vista on Nature. We highlight how those new astronomical messengers are already unveiling the properties of the most extreme astrophysical objects in the universe, probing dynamical space-time and fundamental physics. The dawn of a new era in gravitational (astro) physics. One of the greatest scientific discoveries of this century took place on September 14, 2015, at 9:50 Coordinated Universal Time (UTC) when a gravitational-wave train launched by two black holes colliding at about one billion four-hundred million light-years away passed by the Advanced Laser Interferometer Gravitational-wave Observatory (LIGO) in Louisiana, causing a variation in the proper distance between mirrors of about one ten-thousandth of a proton's diameter. The event was then recorded at the twin LIGO in Washington about 7 msec afterward. The event was dubbed GW150914 (see figure 4.1).[1] A second black-hole coalescence was observed on December 26, 2015 (GW151226),[2] and a third candidate event, LVT151012, was also recorded, but the statistical significance was not high enough to claim a detection.[3] The campaign to record the skies that had begun a half century before was successfully concluded, ushering in the new era of gravitational-wave physics and astrophysics.

GW150914 was an unexpectedly loud event, detected with a signal-to-noise ratio (SNR) of 24. It was initially identified by an (online) generic-transient search, which uses minimal assumptions about waveforms. The highest statistical confidence was obtained with the (offline) optimal matched-filtering searches that employ waveforms as predicted by Einstein's theory of general relativity. By contrast, matched-filtering searches were essential for detecting GW151226, which was an event quieter than GW150914, having a SNR of ~ 13 and an energy spread over about 1 sec (~ 55 GW cycles), instead of 0.2 sec (~ 10 GW cycles).

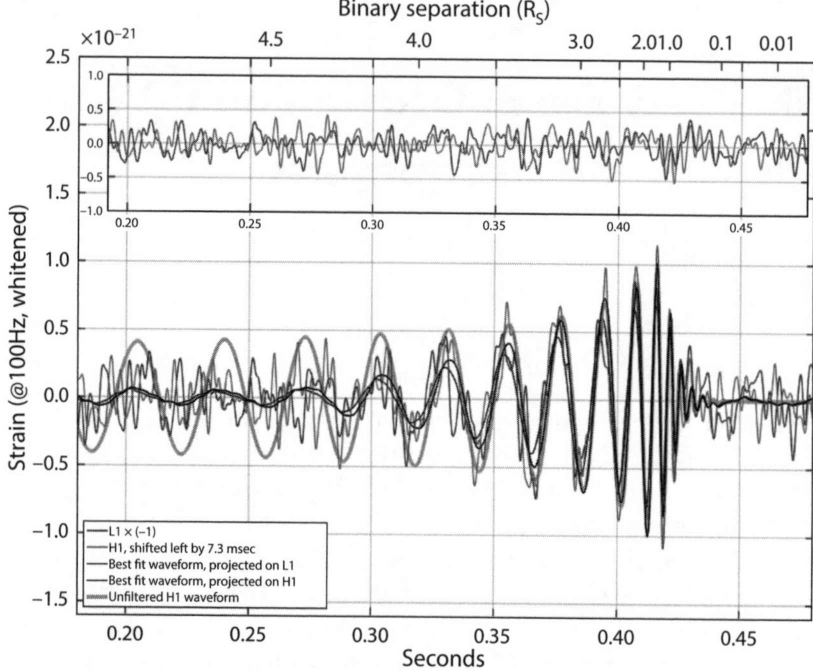

FIGURE 4.1. The gravitational-wave event GW150914 observed by Advanced LIGO's Livingston, L1 (green) and Hanford, H1 (red) detectors, also showing best-fit templates computed by combining analytical and numerical relativity. Data for L1 are shifted by 7.3 ms to account for the time of travel between detectors. Adapted from *2016 Physical Review Letters* 116 061102. See color plate 20.

Black holes are the simplest physical objects in the universe. They are made of warped space and time, no matter. A black hole is fully described by its mass and intrinsic rotation (or spin). The gravitational-wave train emitted by coalescing binary black holes comprises three main stages: (i) the long *inspiral* phase, where gravitational waves slowly and steadily drain the energy and angular momentum from the orbiting black-hole pair, (ii) the *plunge* and *merger*, where black holes move at almost the speed of light, causing a fast variation of the binary orbital period, and then coalesce into the newly formed black hole, and (iii) the *ringdown* stage, where the remnant black hole rings the space-time until it settles down in a stationary configuration (see figure 4.2). Each dynamical stage contains fingerprints of the astrophysical source, which can be identified by first tracking the phase and amplitude of the gravitational-wave train as it sweeps in the detector's bandwidth, and then by comparing it with highly accurate predictions from the theory of gravity.

FIGURE 4.2. Snapshots of numerical-relativity simulation of the binary black-hole coalescence of the gravitational-wave event GW151226. S. Ossokine, A. Buonanno, T. Dietrich, R. Haas (Max Planck Institute for Gravitational Physics), Simulating eXtreme Spacetimes project. See color plate 21.

To detect gravitational-wave signals from coalescing binaries and infer the source properties, LIGO employs waveform models built by combining analytical and numerical relativity (see figure 4.3).[4] The long inspiral phase is characterized by weak gravitational fields and low velocities and it is well described by the post-Newtonian formalism, which expands the Einstein field equation and the gravitational radiation in powers of v/c, v being the binary's characteristic velocity and c the speed of light (or equivalently in $GM/(rc^2)$ with M the binary's total mass and r the radial separation). This approximation loses accuracy as the two bodies come closer and closer to each other, approaching merger. Numerical relativity solves the Einstein equations on high-performance computer clusters and provides the most accurate solution for the last stages of inspiral, plunge, merger, and ringdown. The latter can also be obtained analytically by linearizing the Einstein equations about the black-hole space-time. However, numerical simulations are time-consuming to produce—for example, the state-of-the-art spectral code of the Simulating eXtreme Spacetimes (SXS) project took three weeks and 20,000 CPU-h to compute the gravitational waveform for the event GW150914, and three months and 70,000 CPU-h for GW151226. A few hundred thousand waveform models were used in the template bank of Advanced LIGO during the first observing run to cover the binary black-hole parameter space of total masses 4–100M_0 and mass ratios up to 99. Millions of waveform models are employed in Bayesian follow-up analyses based on Markov Chain Monte Carlo algorithms to infer the properties of the sources and test general relativity. Thus, novel approaches to the two-body problem had to be developed to provide LIGO with accurate and efficient waveform models. The effective one-body formalism builds on and improves existing analytic approximation methods, pushing their validity up to merger. It uses physical arguments deduced from the transition inspiral to merger to ringdown in the much simpler but similar problem of a test body spiraling into a black hole and provides a semi-analytic waveform for the entire coalescence process. Effective one-body waveforms are further improved and completed, achieving the high accuracy required by LIGO analyses by calibrating them to a discrete number of numerical-relativity simulations and extrapolating them to the entire binary parameter space. Fast to generate, quite efficient phenomenological waveform models are also built in the frequency domain by combining effective one-body waveforms at low frequency and numerical-relativity ones at high frequency.

Thus, several decades of patient, steady work in solving the two-body problem in general relativity have finally paid off, although more theoretical work will be needed to solve the new challenges posed by future, more sensitive searches if we want to take full advantage of the discovery potential in the new era of gravitational-wave astronomy.

Astrophysics Implications from LIGO's First Black Holes

The two gravitational-wave signals observed by LIGO, GW150914 and GW151226, have different lengths and strengths, and reach the luminosity peak at different frequencies, about 200 Hz and 500 Hz, respectively. These morphological differences have revealed quite distinct binary black-hole sources. Employing a sophisticated Bayesian analysis that uses multidimensional waveform models, it was found that GW150914 was composed of two stellar black holes with masses $36M_0$ and $29M_0$, which formed a black hole of $62M_0$, rotating at almost 70% of its maximal rotation speed.[5] By contrast, GW151226 had lower black-hole masses, $14M_0$ and $8M_0$, and merged in a $21M_0$ black-hole remnant. Quite interestingly, because the merger signal was less visible for GW151226 (it happened at higher frequency where LIGO sensitivity degrades), it was not possible to measure the binary total mass accurately, and, as a consequence, the binary's individual masses for GW151226 have larger uncertainties (about 30–60%) than for GW150914 (about 10–20%). (What is best measured in a binary coalescence is the chirp mass, which is a combination of the binary total mass and mass ratio. Thus, to have a good measurement of the individual masses, one needs to break the degeneracy and extract the total mass.) Nevertheless, the analysis excluded that the lower mass object in GW151226 was a neutron star, being the component mass larger than $5.6M_0$ at 90% confidence level. (The expected maximum neutron-star mass is $3M_0$.) The follow-up analysis also revealed that at 90% confidence level the individual black holes had spins less than 70% of the maximal value, and that at 99% confidence level at least one of the black holes in GW151226 was rotating at a rotation speed larger than 20% of the maximal value. Because of the signal's length, the orientation of the binary plane (almost face off / on) and the similar orientation of the two LIGO detectors, no information about spin precession could be extracted. (If the spins of the black holes are misaligned with the direction of the binary's orbital angular momentum, they precess on timescales longer than the orbital period,

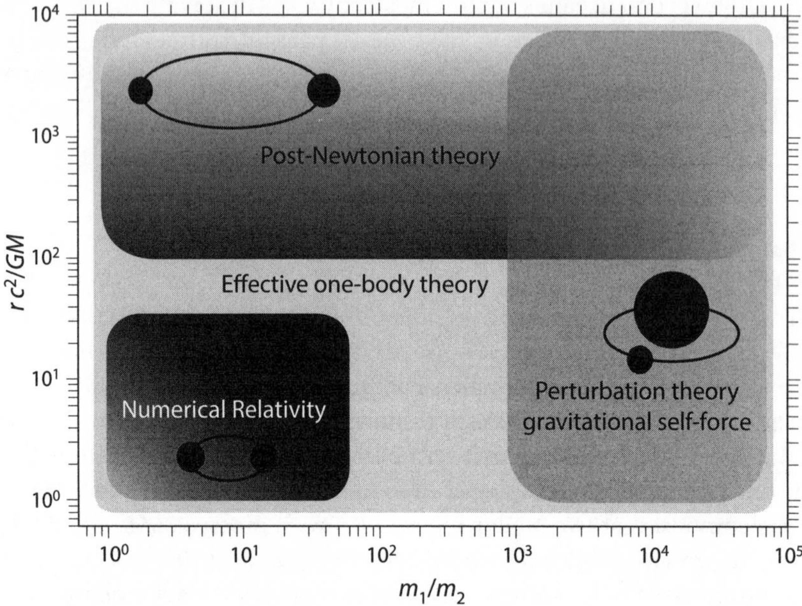

FIGURE 4.3. Current range of validity of main analytical and numerical methods to solve the two-body problem. The horizontal axis shows the binary mass ratio, while the vertical axis shows the radial separation between the two black holes in the binary normalized to GM/c^2, where $M = m_1 + m_2$. Adapted from arXiv:14107832.

inducing characteristic modulations to the gravitational-wave signal.) Finally, the binaries that have produced GW150914 and GW151226 were at comparable luminosity distances from the Earth, about 420 Mpc (i.e., a redshift of 0.1) and the peak of the gravitational-wave luminosity was about 3×10^{56} erg/sec, making them by far the most energetic, transient events in the universe.

There exist two main channels for the formation of binary stellar-mass black holes. In the first one, massive binary stars undergo core collapse at different stages of their evolution, leaving behind a binary black hole. In the second channel, the binary black hole forms by dynamical capture in high stellar density environments, such as globular clusters, typically found in the halo of a galaxy. The former channel will likely predict black-hole spins almost aligned with the orbital angular momentum, while the latter would favor randomly oriented spins. Since precessional effects were not measured, current observations cannot discriminate between or rule out any of the two channels. The observed black-hole masses are consistent with stellar-mass black holes, but for GW150914 they are larger than the maximum black-hole mass observed through X-ray observations. The larger black-hole masses for GW150914 (i.e., 30–40M_0) suggest a different binary population than GW151226, having progenitor massive stars in low metallicity (i.e., tenths of solar metallicity) environments which prevent stars to lose mass because of strong winds before collapsing to black holes.[6] (We note that here metallicity means elements heavier than hydrogen and helium.) Furthermore, whereas the predictions of binary black-hole mergers from astrophysical formation mechanisms varied by several orders of magnitude before LIGO detections, the latter have now established the somewhat high rate of 9–240 $(\text{Gpc})^3$/yr at 95% confidence level. (As a useful comparison, the supernovae rate, which is considered quite high, is about 10^5 $(\text{Gpc})^3$/yr.) Larger black-hole masses and higher coalescence rates point out the interesting possibility that a gravitational-wave stochastic background composed of unresolved signals from binary black-hole mergers could be observed when Advanced LIGO reaches design sensitivity in 2022.

The sky localization of GW150914 and GW151226, which is mainly determined by recording the time delays of the signals arriving at the interferometers, extended over several hundred square degrees. (To put this in perspective, the full Moon seen from Earth is about 0.5 degrees in diameter.) This makes it very difficult to search for an electromagnetic counterpart. Nevertheless, the first campaign for possible electromagnetic counterparts of gravitational-wave signals involved almost twenty astronomical facilities spanning gamma-ray, X-ray, optical, infrared, and radio satellites and telescopes, and culminated with no (convincing) evidence of electromagnetic signals emitted by GW150914 and GW151226. The outcome is not surprising since standard scenarios predict that none (or a negligible amount) of the energy emitted by the coalescence of stellar-mass black holes comes out in the electromagnetic channel. It is expected that at merger, stellar-mass black holes have lost all the gas and matter that were originally present at formation. Astrophysical black holes are also supposed to be neutral since, due to the weakness of the gravitational force, they can only accrete a negligible amount of charge and any charge is quickly neutralized by the opposite charge present in the environment. Deviations from the standard scenario may rise if one considers dark electromagnetic sectors, spinning black holes with strong magnetic fields, which need to be sustained until merger, and black holes surrounded by clouds of axions. Future observations may of course reveal unexpected surprises.

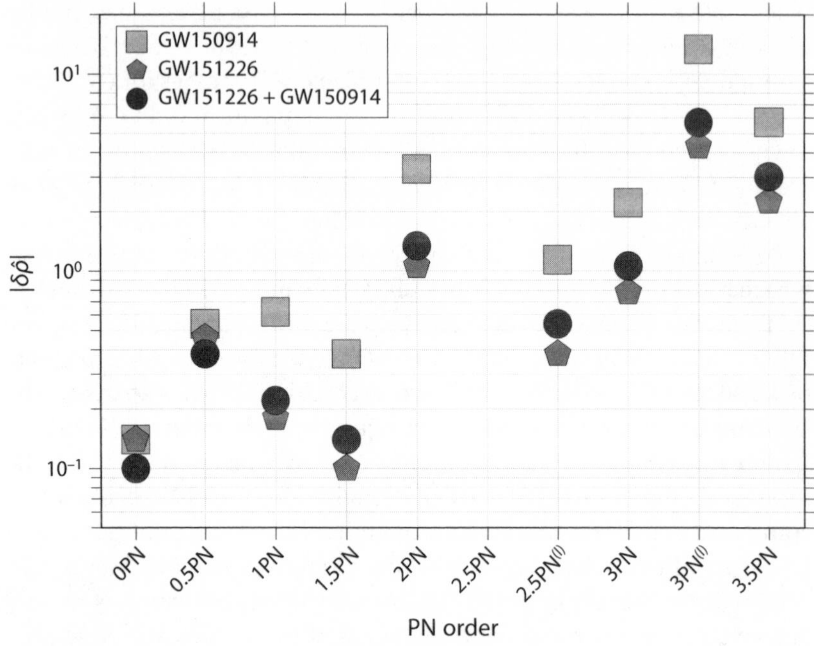

FIGURE 4.4. The 90%-credible upper bounds on deviations in the post-Newtonian coefficients from GW150914 and GW151226, and the joint upper bounds from the two detections. Adapted from 2016 *Physical Review Letters* X 6 041015.

Fundamental Physics Implications from LIGO's First Black Holes

LIGO observations offered the possibility to carry out tests of the theory of general relativity in the so far unexplored, highly dynamical, and strong-field gravity regime. As the two black holes that emitted GW150914 and GW151226 ended their cosmic dance, merging with each other, the binary's orbital period varied considerably. Since the (dominant) gravitational-wave frequency is twice the orbital frequency, the phase of the gravitational-wave signal also changed considerably toward merger. By solving the Einstein equations approximately, it is possible to obtain an analytical representation of the phase evolution in post-Newtonian theory as an expansion in v/c. The coefficients in this post-Newtonian expansion assume specific values in different theories of gravity. Thus, by tracking the phase evolution accurately and comparing it with predictions from the theory of general relativity, it is possible to look for deviations and, if no deviations are found, set upper bounds on those coefficients. The latter describe a plethora of dynamical and radiative physical effects—for example tails of radiation due to backscattering of the gravitational waves with the curved space-time geometry around the holes, spin-orbit, and spin-spin couplings. Whereas long-term timing observations of binary pulsars have allowed us to constrain the leading-order post-Newtonian coefficient in the orbital period derivative with exquisite precision, LIGO observations have obtained the most stringent bounds on higher post-Newtonian terms, setting upper bounds as low as 10% for some coefficients when combining the results from GW150914 and GW151226 (see figure 4.4).[7] Furthermore, using a phenomenological, analytic representation of the merger and ringdown signal, it was also possible to investigate possible deviations during the nonperturbative coalescence phase. In this regime, as well, Einstein theory passed the test with flying colors. Quite interestingly, one could also bound the viscosity that the putative exotic matter of the remnant could have and still be consistent with observations if the compact object formed after merger was not a black hole, but an exotic object such as a boson star, or a gravastar, for example.

Can we conclude that the compact objects that produced GW150914 and GW151226 were black holes (i.e., objects described by mathematical solutions to Einstein equations having a horizon from which neither light nor anything else can escape)? Black holes ring space-time when perturbed. They emit gravitational damped sinusoids or quasi-normal modes whose frequencies and decay times depend only on the black-hole mass and spin (and not on the nature of the perturbation). By measuring multiple quasi-normal modes, it is possible to test if the compact object is described by the black-hole solution in general relativity and extract the black-hole mass and spin. However, the SNR of the two LIGO observations did not allow a measurement of the multipole modes in the ringdown stage. The follow-up analysis for the louder signal, GW150914, could only set upper bounds on the frequency and decay time of one, single mode, finding consistency with the (dominant) quasi-normal mode as predicted by general relativity when using the final mass and spin of the remnant. The latter were derived from the LIGO data by combining the posterior distributions of the component masses and spins with formulae obtained in numerical-relativity simulations of binary black holes in the theory of general relativity. Thus, although the first LIGO observations could neither test the second law of black-hole mechanics, which states that the black-hole entropy cannot decrease, nor the no-hair theorem, which says that a black hole is only described by mass and spin, we expect that future, multiple gravitational-wave detections at higher SNR will shed light on those important theoretical questions.

Despite those limitations, LIGO observations provided the most convincing evidence to date that stellar-mass compact objects in our universe with masses larger than ~ $5M_0$ are described by black holes, that is, by the solutions to Einstein field equations as found by Schwarzschild in 1916 in the spherically symmetric case, and by Kerr in 1963 in the axially symmetric, rotating case.

Future Observations of Coalescing Neutron Stars and Other Gravitational-Wave Signals

During the first observing run, from mid-September 2015 to mid-January 2016, LIGO did not detect gravitational waves from binaries composed of either two neutron stars, or a black hole and a neutron star. Nevertheless, it could set the most stringent upper limits on the rates, notably 12.6×10^3 $(\text{Gpc})^3$/yr for the former and 3.6×10^3 $(\text{Gpc})^3$/yr for the latter at 90% confidence level.[8]

When compared to astrophysical rates based on binary pulsar observations and population synthesis calculations, the LIGO rates imply that we expect to detect those binary systems as Advanced LIGO and Virgo approach and reach design sensitivity in 2022.[9] Assuming a rate for short gamma-ray bursts (GRBs) of 3–30 $(\text{Gpc})^3$/yr and that all GRBs have binary systems as progenitors, the analysis of LIGO data also led to a lower bound on the beaming angle of 1.2–4.0 deg for binary neutron stars and of 2.4–7.4 deg for binaries composed of a neutron star and a black hole. Those lower bounds start to become astrophysically interesting considering that the observed GRB beaming angles vary in the range of 3–25 deg.

The two most exciting consequences of observing gravitational waves from binaries with matter stem from the possibility of inferring the neutron-star equation of state and unveiling the possible origin of GRBs, as we shall now discuss. Neutron stars are extremely dense objects that form when massive stars run out of nuclear fuel and collapse. Hitherto, we do not know what the core of neutron stars is made of. In the core, the neutron star's density is expected to be higher than twice nuclear saturation density 2.8×10^{14} g/cm^3. (For reference, the density of the Sun is 1 g/cm^3.) At those high densities, the standard structure of nuclear matter breaks apart and new phases of matter may appear, including a superfluid quark-gluon plasma. All mass and spin parameters being equal, the gravitational-wave train emitted by a binary containing a neutron star differs from the one emitted by two black holes only in the late inspiral, say 20–30 gravitational-wave cycles before merger, when finite-size effects show up in the waveform because the neutron star is tidally deformed or disrupted due to the tidal field exerted by the companion. Notably, the gravity gradient (or tides) imposed by the companion across the neutron star causes it to deform from sphericity. Depending on the neutron-star equation of state and the nature of the companion (i.e., a neutron star or a black hole), the gravitational-wave train can start to differ from that of binaries consisting of two black holes at frequencies as low as 400 Hz. By tracking the gravitational-wave phase as the waveform sweeps in the detector bandwidth, it will be possible to measure the tidal deformability parameter, which contains information on the neutron-star interior, similar to the parameter measured for Saturn's moon Titan which pointed out the possible existence of a subsurface ocean. With several tens of detections, Advanced LIGO will be able to discriminate some equations of state (e.g., stiff from soft equations of state, where the former allows a neutron star to have a radius as large as 15 km, while the latter up to 9 km). Although Advanced LIGO will not be able to observe the merger signal from binary neutron stars, because the latter occurs at high frequencies, about 1kHz, the use of squeezed light will improve the sensitivity at high frequency and allow to probe the merger signal, measuring neutron star's oscillations and inferring several astrophysical characteristics of the remnant. Squeezed light will be employed in future upgrades of Advanced LIGO and Virgo, and / or next-generation detectors on the ground which are currently under investigation, such as the Einstein Telescope and Cosmic Explorer. The merger of double neutron stars and / or black hole–neutron-star binaries are currently considered the most likely progenitors of short-hard GRBs, which are among the brightest electromagnetic events known to occur in the universe. We expect a plethora of electromagnetic signals from the coalescence of such compact objects: prompt, beamed gamma-ray and X-ray emissions, isotropic optical and infrared after-glows and radio emission. Identifying the electromagnetic counterparts to LIGO detections will demonstrate the short-hard GRB/binary merger paradigm.[10] As discussed above, during the first observing run, the LIGO Scientific Collaboration worked closely with several astrophysicists who used astronomical facilities to follow up the first detections. However, besides the fact that GW150914 and GW151226 were emitted by double black holes, the two events were poorly localized (several hundreds of degree squared), making it extremely challenging to find the *shining needle in the haystack*! Source localization will continue to improve over time, first when Advanced Virgo came online in the Summer 2017, then when the network composed by Advanced LIGOs and Virgo will reach design sensitivity in 2022, at which point we expect that 20% of events per year will be localized within 20 deg^2 at 90% confidence level. The addition of a fourth detector, LIGO in India (2025) or KAGRA in Japan, will allow for even better source localization over the whole sky. Meanwhile, numerical relativists are producing more sophisticated simulations of binary coalescence to predict more quantitatively which equations of state, masses, and spins of double neutron stars and / or black-hole–neutron-star binaries are required to create the astrophysical conditions to produce those electromagnetic signals. Whereas gamma-ray and X-ray signals will be more rare to observe because they require the Earth being along the line-of-sight of the beamed electromagnetic signal, optical / infrared transients known as kilonovae are more likely to be identified because they are isotropic. The kilonova signal is expected to be produced during the post-merger stage by the radioactive decay of unstable nuclei that are produced by the neutron-rich ejecta undergoing rapid-neutron capture (r-process) nucleosynthesis. (Quite interestingly those recent studies have also shown that r-processes ignited by neutron-star mergers can also account for several rare heavy elements in our galaxy, such as gold and platinum.)

Furthermore, at some point during its life, the core of a massive star ceases to generate energy from nuclear fusion and undergoes a sudden collapse (or supernova explosion), forming a neutron star or a black hole, depending on the star's mass. Gravitational waves in the form of unshaped bursts are produced during such a catastrophic event, lasting for tens of milliseconds. At design sensitivity Advanced LIGO and Virgo could detect GW bursts from the core's bounce only if the supernova took place in the Milky Way or neighboring galaxies, but gravitational waves produced in more extreme emission scenarios as predicted by current numerical simulations could be observed up to 10 Mpc.

Pulsars, which are highly magnetized, rotating neutron stars, are also promising astrophysical sources of gravitational waves. Mountains on the crust of pulsars of about a few centimeters in height can cause a variation in time of the pulsar's quadrupole moment, producing a continuous gravitational-wave train at twice the rotation frequency of the pulsar, if the latter rotates along one of its principal axes. The most recent LIGO all-sky searches and targeted observations of known pulsars, such as Vela, Crab, and Cassiopeia A, have started to invade the parameter space of astrophysical interest, setting the most stringent upper limits on the source's ellipticity, which measures the amount of pulsar deformation and depends on the neutron star's equation of state. The most recent all-sky search supported by the computer power of Einstein@Home (a volunteer computer network) has set upper limits for the ellipticity as low as 10^{-6} at 90% confidence level at frequencies of a few hundreds of Hz for pulsars at distances of 100 pc.

Last, several physical mechanisms in the early universe could have produced gravitational waves, such as cosmic inflation, first-order phase transitions, vibrations of fundamental and/or cosmic strings. Gravitational waves being (almost) unaffected by matter and energy encountered while traveling toward the Earth, they provide us with a pristine snapshot of the source and the universe at the time they were produced. Thus, gravitational waves may unveil a period in the history of the universe, around its birth, that we cannot otherwise access. The first observing run of Advanced LIGO has set the most stringent constraints on the stochastic gravitational-wave background, which is generally expressed by the dimensionless energy density in gravitational waves, of $\Omega_{GW} < 1.7 \times 10^{-7}$ at 95% confidence level in the range of frequency of 20–86 Hz for zero spectral index. Digging deeper, at design sensitivity Advanced LIGO is expected to reach Ω_{GW} 10^{-9} and next-generation detectors, such as the Einstein Telescope and Cosmic Explorer, may achieve values as low as 10^{-13}, only two orders of magnitude above the background predicted by the standard slow-roll inflationary scenario if the tensor-to-scalar ratio is $r = 0.11$. (The latter is the current upper limit set by the Planck satellite when analyzing the cosmic microwave background radiation.) However, more exotic inflationary scenarios or the presence of cosmological phases stiffer than radiation at the end of inflation can lead to higher stochastic backgrounds at frequencies probed by LIGO and Virgo. As is customary in science, it is better to keep an open mind and wait for the data to tell us the true story of our universe.

Looking Ahead into the Bright Future of Gravitational-Wave Astronomy

The year 2016 was the best for gravitational-wave research, not only on the ground but also in space. In June 2016, the results of LISA Pathfinder were published.[11] The mission aimed at testing the core technology of the Laser Interferometer Space Antenna (LISA), which is planned to be launched by the European Space Agency as L3 mission in 2034. The goal of the mission was to control and measure with unprecedented accuracy the motion of two test masses in almost perfect gravitational free fall at a distance of about 40 cm inside a drag-free system at the Lagrange L1 point of the Earth-Sun system. The two test bodies are cubes made of gold-platinum of about 2 kg and size of 4.5 cm. The published results exceeded any expectation. The square root of the power spectral density of the relative acceleration noise between the two test masses reached 0.54×10^{-15}g / Hz at frequencies between 0.5–20 mHz, g being the conventional value of the gravitational acceleration. The measurement is more than a factor 5 better than the LISA Pathfinder requirement, and only a factor 1.25 above the LISA requirement. Rightly, LISA Pathfinder has been dubbed the stillest place in the universe! LISA will be sensitive to gravitational waves between 10^{-4} Hz and 10^{-2} Hz, thus detecting sources different from the ones observed on Earth, notably supermassive binary black holes, extreme mass-ratio inspirals composed of a stellar-mass compact object spiraling into a supermassive black hole, and the astrophysical stochastic background produced by double white-dwarf binaries in our galaxy. Quite interestingly, binary black holes of 30–40M_0 or heavier (i.e., the kind of sources detected by LIGO) would also be observed by LISA for a few years, but days or months in advance of LIGO and Virgo depending on the binary parameters.

By searching for correlated signatures in the pulse's arrival times of the most stable, known pulsars, the Pulsar Timing Array (PTA) projects aimed at detecting the stochastic gravitational-wave background.

The sensitivity of existing detectors on Earth will be improved in the next five years using current facilities by employing squeezed light, which allows for decrease of the quantum-optical noise at frequencies above 100 Hz without increasing the laser power. (Squeezing has been already successfully tested in LIGO and GEO600 in Germany.) This improvement in sensitivity will lead to a reduction of the sky-localization errors of coalescing binaries and a better measurement of tidal effects and neutron-star equation of state in binary mergers, and will enhance the chances to observe gravitational waves from pulsars and supernovae. However, to maximize the scientific potential of gravitational-wave physics and astrophysics, a new facility will be required. Several design targets for the next generation of gravitational-wave observatories, to be built in 10–15 years, are currently under study, among them the Einstein

Telescope in Europe and the Cosmic Explorer in the United States. Those future detectors aim at improving the noise spectral density of Advanced LIGO and Virgo by one or even two orders of magnitude depending on the frequency bandwidth. They will allow us to observe binary coalescences with high SNR, > 20, even at high redshift, $z > 10$, and SNR> 100 and $z < 2$, thus probing binaries of stellar- and/or intermediate-mass black holes in all our universe! These observatories will give us the ability to carry out the most exquisite tests of general relativity in the highly dynamical, strong field regime, challenging our current knowledge of gravity, fundamental and nuclear physics, unveiling the properties and the nature of the most extreme objects in our universe.

5

The Wagers of Science

DANIEL KENNEFICK *

The successful detection of gravitational waves by the two LIGO instruments in late 2015 has become, since the announcement of the result in early 2016,[1] one of the most celebrated scientific discoveries of the twenty-first century. A striking, and widely reported, feature of the announcement was the waveform of the event which (after suitable signal processing) showed the signal standing out clearly from the detector noise with characteristic, though short, late inspiral and merger phases (Abbott et al. 2016). What contributed significantly to the convincing case that this signal really represented gravitational waves was the close agreement between the signal and the accompanying theoretically derived waveform which tracked it in the publicly released figure.[2] This figure, unlike most scientific graphs, was not confined to the technical journal article, but features prominently in the public announcement and press releases and was seen by many in the general public. This chapter provides an overview of the story of how that figure came to be made.

In 1998, one of the founders of the LIGO laboratory, Kip Thorne, challenged his fellow theorists to a wager. He predicted, correctly, that the first gravitational waves detected by LIGO would come from a coalescing black hole binary. He bet a roomful of numerical relativists, theorists who attempt to solve Einstein's equations for gravity using giant programs running on supercomputers, that LIGO would have recorded such signals before they had been successful in predicting the waveforms using their codes. The numerical relativists took the bet, but voices could be heard asking, "what are the odds?," "what's the spread?," and some of those trooping up to sign the wager, which Kip had ready for them in paper form, did not, in my judgment, look that enthused to be taking the wager.[3] At any rate the wager was made, and cheerfully conceded by Kip in 2016. The background to this event, rather than the breakthroughs made subsequently which allowed the bet to be won, will be my focus in this chapter.

A History of Controversy

One key feature of the background to this wager is the historical controversy over whether gravitational waves exist at all or are predicted by the theory of general relativity (GR) to be emitted by binary systems such as the source of GW150914. It will therefore be helpful to give a brief account of these controversies.[4] Skepticism over the existence of gravitational waves was born with the theory of GR, one of the first skeptics being the theory's discoverer, Einstein himself. In February 1916, he wrote to Karl Schwarzschild that he did not think that his new theory predicted the existence of gravitational waves (Einstein 1998, Doc. 194).[5] Within a few months Einstein had changed his mind, publishing a paper which discussed them at length in the context of the linearized approximation to his theory (Einstein 1916). This paper contained a serious error that, when he discovered it, obliged him to completely rewrite it, resulting in a new paper titled "On Gravitational Waves," published early in 1918 (Einstein 1918). The new paper introduced the quadrupole formula giving the flux of energy emitted by a source of gravitational waves and it constitutes the first concrete theory of the phenomenon.

Reflecting the difficulty of identifying gravitational waves in an unambiguous way, in his 1918 paper Einstein discussed three different types of gravitational waves, two of which turned out to be spurious. They were merely flat space seen in a "wavy" sort of coordinate system. In 1922, Arthur Stanley Eddington wrote a paper intended to show that gravitational waves must travel at the speed of light (Eddington 1922). Indeed, he confirmed this for the one real type of wave discussed by Einstein. He realized, however, that the spurious waves could travel at any speed, depending on the coordinate choice of the person doing the calculation. In this way he realized that these two types of waves were spurious. As he put it, they travel not at the speed of light, but at the "speed of thought."

After 1918, Einstein left the topic of gravitational waves alone for nearly two decades before returning to it in collaboration with his first American assistant, Nathan Rosen. In trying to find a solution representing gravitational waves in the full theory of GR (not in the linearized approximation to it), they encountered a problem with singularities that kept appearing in the metrics they were discussing. They constructed a proof that it was impossible to find a solution for plane waves without such a singularity and submitted a paper to the *Physical Review* entitled "Do Gravitational Waves Exist?" The answer they gave was no. The paper was reviewed for the journal by another Princeton physicist, Howard Percy Robertson, who was then (in mid-1936) on sabbatical at his alma mater, Caltech. Robertson argued against Einstein and Rosen's conclusions, and his ten-page anonymous report was returned to Einstein by the *Review*'s editor, John Tate, with the gentle request that Einstein might like to respond to the referee before the paper was published. Einstein replied in high dudgeon that "he had sent the paper for *publication*" and withdrew it in light of the "erroneous" criticisms of Tate's "anonymous expert." He never published an article in the *Review* again (see Kennefick 2005 or 2007 for the full story).

In the fall semester of 1936, Einstein was joined by a new assistant, Leopold Infeld from Poland, who replaced Rosen. Infeld struck up a friendship with Robertson, who had returned to Princeton from his sabbatical. Taking advantage of this new entre into Einstein's group, Robertson persuaded Infeld that Einstein's gravitational-wave argument was wrong. He never revealed to Einstein, or even to Infeld, that he had, in fact, refereed the paper. Infeld, in his autobiography (Infeld 1941), charmingly marvels at the speed and surety with which Robertson had found the error in his arguments one afternoon, little realizing that Robertson was quite familiar with Einstein's still unpublished paper, having spent a week poring over it "for the sake of my own soul," as he commented to Tate (Kennefick 2005). Einstein altered the paper in proofs, changing everything, up to and including the title, which became "On Gravitational Waves" (Einstein and Rosen 1937). The problem of the singularity had been finessed by altering the geometry of the space-time from planar to cylindrical symmetry so that the singularity could be associated with a possible source along the central axis. Only after the war was it shown that the singularities that so bothered Einstein and Rosen are, in fact, coordinate in nature and pose no threat to the existence of gravitational waves (Bondi, Pirani, and Robinson 1959).

Skepticism of the existence of gravitational waves as a prediction of the theory did not end with Einstein's reversal on the issue in 1936. Both of Einstein's assistants of the period, Rosen and Infeld, went on to champion the cause of skepticism in the postwar period. Rosen tried to argue, in 1955, that gravitational waves did not carry energy with them as they propagated and were therefore more spurious than real (Rosen 1955). This argument was addressed at the Chapel Hill conference of 1957, when Richard Feynman and Hermann Bondi, drawing on the work of Felix Pirani presented at that meeting, produced the famous "sticky-bead" thought experiment to show that gravitational waves must interact with matter in such a way as to transfer energy to material systems (De Witt 1957; Bondi 1957; see Kennefick 2007 for a thorough discussion). This argument was generally taken to have answered Rosen's objection.

Infeld's point, that gravitational waves were real enough, but were not emitted by binary star systems (such as the system recently detected by LIGO), was regarded by Bondi as a legitimate one. In trying to answer it, Bondi and his group at King's College, London (which also included Pirani and which collaborated with a student of Infeld's, Andrzej Trautman and several other visitors, such as Roger Penrose, Ted Newman, and Rai Sachs) developed many key ideas, such as the Bondi news function, which demonstrated that gravitational waves did indeed carry away energy from binary systems (Bondi, van der Burg, and Metzner 1962; see Kennefick 2007 for many other relevant references from this period). This work was largely complete by the mid-1960s, but the quadrupole formula controversy continued into the 1970s as some experts, such as Peter Havas and Juergen Ehlers, were unhappy with the level of rigor used to show that this formula could be legitimately applied to binary sources. The fact that binary systems lost energy to the emission of gravitational waves having been admitted, the question was *how much* energy was emitted. Did Einstein's quadrupole formula apply to such strong field systems as binary black holes (Kennefick 2007)?

This controversy was overtaken by the discovery of the first binary pulsar system in late 1974 by Joe Taylor and Russel Hulse (Hulse and Taylor 1975). Taylor and collaborators demonstrated that this system, consisting of two neutron stars in an extremely close orbit with a "year" of only eight hours, was decaying in its orbit by an amount in agreement with the prediction of the quadrupole formula (Weisberg and Taylor 1981). This was the first evidence for the existence of gravitational waves. Not only did it win the Nobel Prize for Taylor and Hulse, but it also made the LIGO project possible by providing the first observational evidence that gravitational waves really existed.

The effort to achieve gravitational-wave detection began as early as 1960 when Joe Weber, who had been a participant in the debates around the time of Chapel Hill, inaugurated the field. In 1969, he announced that he was detecting gravitational waves (Weber 1969), giving birth to a controversy that inspired many new groups to try to detect gravitational waves. The controversy ended in the conclusion that Weber had not seen them (Collins 2004). Perhaps surprisingly, in hindsight, the Weber controversy did not discredit the field, and instead the confirmation by Hulse and Taylor that significant sources of gravitational radiation existed gave renewed optimism to those who, inspired by Weber's example, hoped for a detection of gravitational waves by Earth-based detectors. Two of those new groups, one based at MIT, the other at Caltech, decided to pursue detection of gravitational waves by interferometric detectors fundamentally different from Weber's bar detectors (for the history of gravitational-wave detection, see Collins 2004; Levin 2016; and Bartusiak 2017; for a close analysis of LIGO's actual operation, read Collins 2014, and for a thorough account of the detection of GW150914, see Collins 2017).

Theorists' Role in LIGO

In 1991, I joined Kip Thorne's research group as a graduate student at Caltech. Kip held group meetings every week at which students and postdocs would discuss their ongoing research and get advice from Kip and others in the group. At one of the first meetings I attended, Kip came in with a large sheaf of printouts of notes he had written that mapped out a range of tasks to which theorists needed to attend in order to help LIGO detect gravitational waves. At this time, efforts to fund LIGO had borne fruit with the first congressional appropriation of funds for the project in 1991. In 1992, the sites for the detectors were chosen (Anonymous 2016). At this point it was clear that LIGO was likely to be built and go into operation, yet Thorne was worried that the theoretical tools to make it possible for LIGO to detect anything did not yet exist. LIGO's signals, including GW150714, are much too weak to be visible over the detector noise. Some elementary processing is sufficient to make the GW150714 signal apparent to the eye compared to the remaining background noise (some noise has been filtered out in published graphs). This is a consequence of the fact that this signal is an unusually strong one, though with only a small number of cycles visible in the signal (the reason for that is that the large mass of this system gives it a low frequency and it is only in the last few cycles of the inspiral that it enters the LIGO waveband). Neutron star binaries, on the other hand, can be expected to be buried much more deeply in the noise. In compensation, many cycles of the inspiral will be present in the LIGO signal, because these lower mass binaries will have higher frequency signals, falling well within the most sensitive part of LIGO's waveband, which peaks in the vicinity of 100 Hz. Thus, using optimal filtering to integrate over the full signal, one can raise the signal-to-noise ratio to reasonable levels and thereby make a detection of such systems.

There are two issues with the optimal filtering approach to signal analysis. One is that this is not really the same as seeing something visible above the noise, so it was perhaps fortunate that the first signal was such a strong one. Second, one cannot do optimal filtering unless one already knows what the signal looks like, because one essentially compares the signal to a template of the expected waveform. Since gravitational waves have never been detected before, it follows that only theory can provide the templates required for this kind of data analysis. But in the early 1990s, no theoretical templates of this kind existed and the purpose of Thorne's notes was to lay out the tasks required to create those predicted waveforms. He informed us graduate students that the group would focus our research on gravitational waves and LIGO data analysis for the foreseeable future, and he encouraged other groups to take an interest in these issues, with some considerable success. Such was his success in getting his own students and postdocs on board that a colleague who joined the group in this period told me that his previous group referred to Thorne's group as "the Juggernaut."

In what follows I'll focus on only a couple of major issues addressed in Thorne's notes. Those interested in the content of those notes should consult "The Last Three Minutes: Issues in Gravitational Wave Measurements of Coalescing Compact Binaries" (Cutler et al. 1993),[6] the paper published by Thorne's group outlining both the content of his original notes and the result of the sustained period of effort by his research group that followed the meeting when the notes were presented to us.[7] The title was chosen to consciously echo Steven Weinberg's famous book about the origins of the Universe (Weinberg 1993). It refers to the fact that it is the last three minutes of the typical inspiral that was expected to fall within LIGO's waveband. So far, interestingly enough, LIGO has mostly been successful in detecting signals that last for less than a second.

One key topic Thorne highlighted at the initial group meeting was the need for theorists to work on data analysis issues for LIGO. This was somewhat unusual. Data analysis is typically seen as experimental work. After all, it is only natural that analysis of the data from an instrument should be performed by those most intimately familiar with the instrument and its operation. Indeed, it was something of a culture shock, for Thorne's students, and later for others in the field, for theorists to find themselves engaged in such work, and many did not find it particularly congenial. Nevertheless, it became increasingly apparent that this presented a means for theorists to gain access to LIGO data in a meaningful way and to make their mark on what would be a major event in the field.

Not only did theorists have to get used to doing this kind of work, but experimenters had to adjust to theorists being involved in data analysis. Rai Weiss, the originator of the LIGO concept, became famous for his verbal lacerations of theorists who made statements about LIGO data analysis of which he disapproved. A key aspect of these complaints would be that theorists were straying too far into the experimenter's domain when they attempted to discuss specific aspects of the detector's noise. This was the domain of the experimenters, who alone understood the quirks of their instruments. Why then, should Weiss and other experimenters accept the theorists' involvement at all? The answer is twofold. First, it was a question of shortage of (wo)manpower. LIGO was, in many ways, built and operated on a shoestring. Experimenters had enough to do to get the detector up and running. That help was needed on the complex issue of signal extraction was recognized by LIGO director Barry Barish when he founded the LIGO Scientific Collaboration (LSC) to draw in scientists, many of them theorists, who were not directly employed by the LIGO laboratory. The lure to attract these scientists would be access to LIGO data. The second major reason is that the signals would have to be matched to theoretical templates in a process known as matched filtering, to extract them from the detector noise. Matching to such templates is commonplace in signal analysis, but usually it is done where one has previous observational experience of what the signals look like. In this case, no such experience was available. Instead, theorists would model the signals and thus provide the templates. Just as the experimenters would understand the detector noise, so the theorists would understand the templates and their quirks, having produced them themselves.

Templates for binary black hole mergers would, as Thorne recognized in his notes, have to come from two different sources. The first was analytical. In order to end the quadrupole formula controversy and to prove that the orbital decay of the binary pulsar really did conform to the predictions of GR, theorists had used so-called post-Newtonian calculations to describe the waves emitted by binary systems. These calculations would need to be carried through to much higher post-Newtonian order to be useful for LIGO data analysis. Responding to Thorne's clarion call, the theoretical community was able to provide these post-Newtonian templates, using and developing tools established over the several decades of previous work in the field.

However, it was also apparent to Kip that post-Newtonian approximations would be invalid for the actual merger phase of the interaction between the two black holes. This is precisely the phase that is encountered in the signal from GW150914. For that, one has to solve the exact equations of general relativity, which can only be done numerically on supercomputers. Although Thorne's own group at Caltech did not then contain any numerical relativists, he began to try to attract experts in this area to Caltech as part of an effort to promote progress in this area. Ultimately Caltech would become one of the players in the production of numerically derived templates for LIGO.

Numerical Relativity

The effort to find numerical solutions of the Einstein equations using computers began in the 1960s. Like many things in general relativity, the seeds were sown at the first of the GR series of conferences held in Chapel Hill, North Carolina in 1957 (de Witt 1957), the same meeting where Feynman presented the sticky-bead argument. At this meeting (today known as GR1), Charles Misner and Bryce de Witt both discussed how one might try to solve the Einstein equations numerically on a computer. The first simulation of two black holes was reported in 1964 by Susan Hahn and Richard Lindquist (Hahn and Lindquist 1964). Lindquist, like Misner, was a student of John Wheeler (and thus a relativist), while Hahn was a young product of the then new field of numerical computing, she was an applied mathematician (Holst et al. 2016). As Hahn worked for IBM, she had access to one of the leading supercomputers of the day, though its computing power was significantly less than a modern smartphone. As such their results were more in the nature of a proof of concept than a solution of the problem. The first successful extraction of gravitational-wave signals from a black hole collision was achieved by Larry Smarr, a student of de Witt's, in 1977, when computers had improved; this also led to improved comprehension of how to tackle the problem (Smarr 1977).

In order for the numerical approach to work, space-time has to be divided into a grid, and the rules of general relativity place restrictions on how that grid must be constructed, since the grid represents space-time itself and general relativity insists that space-time's geometry is conditioned by the masses placed within it. The grid cannot be conceived as an inert background against which the physics can be modeled. In general relativity, space-time is an active participant in the physics and is molded by it. Most experts agree that progress was slow during the

1980s and 1990s, and a significant problem was access to supercomputers, which prompted Smarr to urge the National Science Foundation (NSF) to begin supporting numerical science.[8] Smarr's initiative led to the foundation of the National Center for Super-Computing Applications (NCSA) in Urbana-Champaign, Illinois. Ultimately the slow progress on the binary black hole problem, and the urgency introduced by the projected needs of LIGO, prompted the NSF to fund the Binary Black Hole Grand Challenge Alliance (BBHGCA) from 1993 to 1998, which brought most of the existing numerical-relativity groups together to tackle this key problem (Lehner,2001; Cardoso et al. 2015).

As an illustration of how difficult the theoretical struggle has been, go back to November 1998, when the final meeting of the BBHGCA was held. This NSF-funded project had brought together more than a half-dozen numerical-relativity groups to try to address precisely the challenge of producing templates of binary black hole coalescences for LIGO. Progress had been slow enough up to that point that no existing code could describe the evolution of a three-dimensional binary black hole system for more than a small fraction of an orbit before the code crashed. This was a long way short of the more than half- dozen cycles shown to the public in the LIGO announcement of 2016. In the final session of the meeting one of LIGO's founders, Kip Thorne, gave a talk summarizing the achievements of the BBHGCA and discussing what remained to be accomplished. In order to highlight the importance to LIGO of the success of the numerical-relativity effort, he challenged the leaders of the various research groups involved in the Alliance to a wager. Thorne presciently predicted that the first signals LIGO would see would be from the coalescences of binary black holes, precisely the systems the Alliance was trying to model. He bet that LIGO would have recorded signals from binary black hole coalescences before numerical relativity waveforms were available to make sense of these signals. I attended this session and recorded it. Most of the numerical relativists did not appear confident as they queued to sign up for the wager. One asked what the odds were and another what the spread was. Yet they took it. They could not back out of it. Today the wager, conceded by Thorne, is framed and will take its place outside his office with a number of very famous scientific wagers with luminaries such as Steven Hawking. It is a monument to the success of numerical relativity. Indeed, this bet had already been proposed by Thorne before 1998, at least to Larry Smarr, the founding father of numerical relativity; in an interview conducted by me in 1997, Smarr referred to this bet as being in existence (at least "roughly speaking") at that time. He also said: "I think he's [Thorne] got an even chance to beat me."

In the meantime, in 2005, after several years of frustration, a breakthrough in the field suddenly occurred (Sperhake 2015), initially through the work of Frans Pretorius (Pretorius 2005) and then, as if in a dam burst, through two other groups (Campanelli et al. 2006; Baker et al. 2006). Pretorius's initial breakthrough had been a qualitative change because he had been able to evolve the virtual black hole binary for more than one complete orbit. Within a few years the community went from that point to routinely being able to evolve multiple orbits. This permitted the creation, in the early part of this decade, of a bank of templates to be used in LIGO data analysis. Meanwhile the initial LIGO detector, which was online and making observations in 2004 and 2005, failed to see anything, so it was not until advanced LIGO was switched on in 2015 that LIGO finally had a signal to report, though that signal came almost immediately after the devices began operating, before the first science run had even officially begun. Thus, after fifty years of effort, dating back to pioneering work in the mid-sixties, the numerical-relativity community had just dipped at the tape to win their race with destiny, as well as their bet with Thorne. How was this accomplished?

As an example of the kinds of issues that dogged the field in the pre-breakthrough years, consider the problem of excising the singularity at the center of the black hole from the computational grid. The singularity, where space-time curvature is infinite, presents a particular problem for numerical computation. Famously, computers crash when invited to consider the infinite. Fortunately, it appears at first glance that the situation is saved by the existence of the event horizon, which prevents any information from inside the black hole from propagating outward through the horizon. Thus, any difficulties with boundary conditions inside the horizon would be quarantined from infecting the numbers in the rest of the grid. Unfortunately, it is only the numbers that precisely obey the Einstein equations that are prevented from crossing the horizon boundary. Numerical error, which exists because of the finite size of the grid's mesh, by definition does not satisfy the equations. It can cross the horizon because it can travel faster than the speed of light. To paraphrase Eddington, numerical error can travel at the speed of thought, and the speed of thought of a supercomputer is blazingly fast. Errors propagate rapidly across the grid, pile up, and crash the code. After years of futility attempting to address issues such as these, several different approaches bore fruit at about the same time in 2005. Considering that it was individuals and small groups publishing separately who made the breakthrough, one has to wonder if there was something misguided about the whole concept of the grand challenge alliance. Could the breakthrough have been made earlier if the "big science" approach had been rejected as being unsuited to the domain of theory?

In an interview conducted in 1997, Larry Smarr, one of the founders of the field of numerical relativity, emphasized to me the importance of building a toolkit for numerical relativity. Although progress in the field had been, in Smarr's view, slow in solving the problem of binary black hole inspirals, still the BBHGCA would prove to be a success if it helped build the tools that would subsequently be used to solve the problem. Smarr emphasized the poverty of tools available to numericists when compared to the wealth of tools available to analytic theorists studying relativity. The latter community had taken decades to build up their toolkit. These tools play a major role in what Thomas Kuhn called "exemplars," or paradigms in the narrow sense. Exemplars are solutions to central problems within a scientific community that can serve as blueprints for the solution of further problems in the discipline. Such exemplars play a key role in the progress of normal science. Although Kuhn's disciplinary matrices, or paradigms in the broad sense, which govern the whole outlook of a field of science, have received more attention, exemplars play a vital role in the sense that progress within a paradigm is impossible without them. Indeed, a young scientific community would typically start with exemplars, which are building blocks in a net of relationships, models, and tools that make up a paradigm in the broad sense. Exemplars, as research-guiding problem solutions, can involve models, techniques, and calculational tricks that scientists use to tackle new problems in familiar ways. Smarr's argument would be that in 1997, numerical relativity was a relatively immature field with an insufficiency of exemplars; this prevented it from making substantial progress on difficult problems. Twenty years later it is instructive to examine how this tool building was done, and which exemplars, as Kuhn would have it, played a key role in making the breakthrough in the problem of binary black holes.

One particular problem that numerical relativists have attempted to solve with toolkit building is the problem they refer to as "Collaborational Infrastructure." The tools used by numerical researchers sometimes have more in common with experimental apparatus than with typical theoretical tools. They are large, complex bodies of code that require many (wo)man hours to construct, and they require constant upkeep and tending to remain functional. Few research sites have the (wo)manpower to maintain these apparatuses. Thus, to continue with the analogy with experimental work, numerical relativity can benefit from a form of big science. Yet the groups remain small, closer in size to typical theoretical research groups. To address this problem, groups in different places must collaborate. Although few collaborations today are as large as the BBHGCA, nevertheless connections between groups are multifarious and complex and are aided by using common bodies of code.

The Cactus code was developed to facilitate the collaborational infrastructure that has developed within the field, and it is one product of the grand challenge alliance. The Cactus code takes its name from the fact that the main code typically does not actually run simulations; instead, different users write their codes, referred to as thorns, which plug in to the main code.[9] Cactus itself can be used in other fields that have nothing to do with gravitational physics (though Cactus was written with this purpose in mind). The period of the GCA was when Cactus was being written and coming into widespread use. Since then it has given birth to spin-off projects, such as Whiskey, which can add Hydrodynamic code to Cactus to permit the modeling of binary neutron stars as well as binary black holes. Binary neutron stars are another expected source for LIGO and are also very difficult to model numerically. Since Cactus is now used for purposes other than relativity, the Einstein Toolkit is the name for the suite of thorns that permits users to simulate general relativistic systems such as binary black holes (and, with Whiskey, neutron stars). Since the history of numerical relativity exhibits this interesting tension between the need for multigroup collaborations and yet the need for individual groups to innovate creatively, it may be that the study of the history of this field will provide interesting insights into the thorny question of the advantages of big science versus little science.

Conclusions

Today the numerical-relativity community is well integrated into the LIGO Scientific Collaboration (LSC). It provides templates for data analysis to aid in the searches through the advanced LIGO data which it is hoped will uncover new signals. It also provides bespoke waveforms to match already identified signals such as the original GW150914. This is important not only because it is the close fit between theory and experiment which provides assurance that a gravitational wave has been detected, but also because it is only in this way that the source of the waves can be identified. Thus, it is from the numerical relativists that we learn that the system that emitted GW150914 consisted of two black holes, one with a mass of thirty-five solar masses, the other with a mass of thirty solar masses. Post-Newtonian templates can also be used in this way, but play little role in the case of high-mass systems such as GW150914 because the low frequency of the gravitational waves from these sources means that only the very last few cycles are visible within the LIGO waveband. These last cycles and the merger phase itself are the domain of numerical relativity.

One particularly interesting, and widely reported, aspect of the signal is known only from the numerical models. The merged black hole is said to have a mass of sixty-two solar masses, which is fully three solar masses less than the combined mass of the original two objects. What happened to this missing mass? It was radiated away in the form of gravitational-wave energy. However, the mass of the merged object can only be inferred from the ringdown phase of the gravitational-wave signal. These are waves emitted by the merged black hole as its event horizon vibrates back and forth before settling down to a natural spheroidal shape. But initially LIGO was not successful in detecting the ringdown phase with a sensitivity that permitted the mass of the merged object to be inferred from the signal. It is the models produced by the numerical relativists that provide the mass of the merged object. For this interesting part of the publicly told story we are still entirely in the domain of theory. In this way, after a century of effort, the theoretical community has finally made its mark on the world of popular science.

It is now hoped that my field will begin a transition from gravitational-wave physics to gravitational-wave astronomy, a process that will also be fascinating to chronicle from the historian's perspective. On a personal and related note, one small event on the day of the announcement of the GW150914 detection summed up the excitement I felt, excitement that was shared by everyone in the field. My wife, an astronomer who studies quasars, has often been amused by terms such as "electromagnetic astronomers," a term she first encountered in reading a draft of Thorne's popular book (Thorne 1995) which he had asked his students, including myself, to read and comment upon. The term is a gravitational-wave physicist's way of referring to traditional astronomers. It envisages a day when gravitational waves, neutrinos, and other non-electromagnetic messengers will play a major role in astronomy. Seeing the term used in

Kip's draft, she retorted (to me) that only when gravitational waves had actually been detected would there be much need for the term "electromagnetic astronomers." Until then she would remain an astronomer, tout court.

A few years ago I appeared on a television show to talk about the history of the 1919 eclipse expedition to test general relativity. Because of the subject matter it discussed, the show introduced me as "Daniel Kennefick, Astronomer." My wife was amused and took to introducing me to colleagues by saying, "he's not a real astronomer, but he does play one on TV." After the press conference announcing LIGO's first detection she turned to me and said, "now, you're a real astronomer." I wasn't a member of the LIGO Scientific Collaboration and played no role whatsoever in making the great event possible, but still, like many other gravitational physicists, I felt my discipline had entered a new era, confirmed by my promotion within the family's ranks.

6

The Genesis and Transformation of General Relativity

JÜRGEN RENN*

The discovery of gravitational waves is one of the most spectacular breakthroughs of physics.[1] How are such breakthroughs achieved? Was the first terrestrial measurement of gravitational waves simply an expected outcome of the persistent efforts that were made to confirm Einstein's prediction a century ago? In the following, I argue that the pathway to this discovery was neither one of linear progress nor of sudden paradigm shifts but rather the result of a long-term transformation of knowledge with many surprising twists and turns.[2]

The 2015 breakthrough has a long prehistory.[3] It goes back to calculations performed by Einstein a century ago and even beyond, to the first discussions of gravitation as a field comparable to the electromagnetic field and to speculations about its propagation with finite speed. Einstein's formulation of special relativity in 1905 was a first major turning point in this long prehistory as it definitively challenged Newton's conception of gravitation as an instantaneous action at a distance.

A Novel Kinematic Framework

Special relativity emerged from a clash between two principles: relativity and the constancy of the speed of light.[4] Reconciling these two principles entailed a conceptual revolution of space and time. It also implied that most famous of all formulas, which describes the equivalence of energy and mass. What implications did this reconciliation of principles have for understanding the phenomenon of gravitation? All classical laws, including Newton's law of gravitation, had to be adapted to the new framework of space and time that was established by special relativity. In 1907, Einstein made a first comprehensive attempt at such a revision in the context of a review article he was asked to write.[5] In preparing this article, he observed that the adaptation of Newton's law confronted him with the question of whether a peculiar property of gravitation from the classical theory could be preserved in the new framework, namely, that all bodies fall with the same acceleration due to the equality of gravitational and inertial mass. Pondering this question gave rise to what he later described as "the happiest thought of my life."[6] This thought came to him while he was still working as a clerk at the patent office in Bern, where he accomplished all of his early breakthroughs.

The basic idea can be illustrated with the help of one of those famous thought experiments that Einstein liked so much.[7] When two people—one on a moving train and one standing on the platform—simultaneously each drop a stone, the question arises: will the two stones hit the ground simultaneously? According to Newton, yes; according to special relativity, no, because two events that are simultaneous in one frame of reference will in general not be simultaneous in a reference frame in relative motion with respect to the first frame. This was not only surprising but actually seemed to violate the universality of free fall familiar from classical physics. Was there any way to impose this classical insight also within the new framework? This question inspired Einstein to base his efforts to formulate a relativistic theory of gravitation on Galileo's principle that all bodies fall with the same acceleration and on the equality of inertial and gravitational mass accounting for this principle in Newtonian physics.

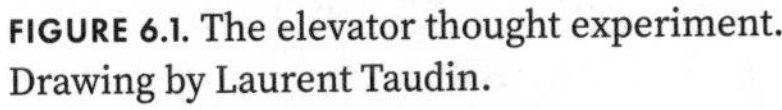
FIGURE 6.1. The elevator thought experiment. Drawing by Laurent Taudin.

The Principle of Equivalence as a Borderline Problem

Einstein was now just a short step away from the idea of simulating the effects of gravitation by imagining an observer in an accelerated laboratory in outer space where bodies fall to the ground because of the inertial forces acting in such a laboratory. Inertial and gravitational forces were thus conceived as being two sides of the same coin, just as electric and magnetic forces can be considered as the two aspects of one electromagnetic field. Which of these aspects one sees in a particular situation depends, in both cases, on the state of motion of the reference system. This frame dependency of the field was a key insight that Einstein was able to take from his work on the electrodynamics of moving bodies (at the roots of special relativity) and now apply to a relativistic field theory of gravitation.

He later recalled:

> At that moment I got the happiest thought of my life in the following form: In an example worth considering, the gravitational field has a relative existence only in a manner similar to the electric field generated by magnetic-electro induction. *Because for an observer in free-fall from the roof of a house, there is during the fall*—at least in his immediate vicinity—*no gravitational field.*[8]

In other words, locally the gravitational field can be transformed away by using a free-falling frame of reference, and a gravitational field can, vice versa, be simulated by the inertial effects occurring in an accelerated frame of reference. In particular, an inertial frame of reference with a homogeneous static gravitational field may be considered as being equivalent to a uniformly accelerated frame of reference with no gravitational field. This is what came to be called Einstein's "principle of equivalence."

In the early years of probing a relativistic theory of gravitation, Einstein made extensive use of this principle in order to explore the implications of a relativistic theory of gravitation, long before he had even attempted to formulate a complete theory. Among these implications were the bending and redshift of light in a gravitational field, as well as a slight rotation of the orbits of the planets in addition to the effects predicted by Newton's theory.

Where did Einstein's revolutionary ideas come from and why did they have such transformative consequences for the architecture of physical knowledge? The answer to this question has two parts, one related to the structure of the shared knowledge of classical physics, the other related to the specific perspective under which Einstein appropriated this knowledge in the course of his early career.

As far as the structure of the shared knowledge of classical physics is concerned, we notice that the trail-blazing ideas of Einstein's miraculous year 1905, the idea of the light quantum, of Brownian motion as a stochastic process, and of the relativity of simultaneity, all emerged from conceptual conflicts within classical physics.[9] More specifically, they originated at the borderlines of different domains of knowledge where different conceptual systems clashed with each other. For instance, his paper on the light quantum emerged from a borderline problem between thermodynamics and electrodynamics. His paper explaining Brownian motion emerged from a borderline problem between mechanics and thermodynamics. And his paper on the electrodynamics of moving bodies, giving rise to special relativity, was triggered by the clash between the relativity principle of mechanics and the principle of the constancy of the speed of light implied by electrodynamics.

After the new space-time framework of special relativity had been established, a further borderline problem raised its head: the challenge to create a relativistic field theory of gravity. This would have to take

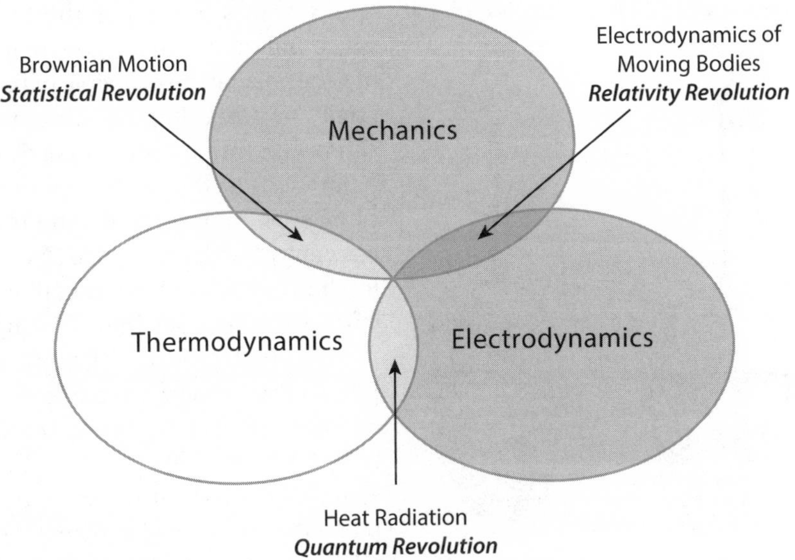

FIGURE 6.2. Revolutions at the borderline problems of classical physics.

into account the law of gravitation of classical mechanics, as well as the characteristics of the field theory first revealed by electrodynamics and then firmly established by special relativity. In tackling this borderline problem and its repercussions on the architecture of knowledge, Einstein created general relativity. So much for its roots in the shared knowledge of classical physics.

Einstein's perspective clearly was very special. Although these borderline problems were in principle visible to anyone who was interested in looking, not everyone perceived them as fundamental challenges. Why was that so? In his 1917 popular book on relativity, written almost immediately after concluding an eight-year struggle with the formulation of general relativity, Einstein wrote:

> Since the introduction of the special principle of relativity has been justified, every intellect which strives after generalization must feel the temptation to venture the step towards the general principle of relativity.[10]

Einstein was in fact one of the few physicists who actually recognized this need. His perspective was shaped by the broad scope of his reading, which included the philosophy and history of science, and in particular by his fascination with Ernst Mach's historical critical analysis of mechanics.[11] There Mach had criticized Newton's concept of absolute space and suggested the possibility of generalizing the principle of relativity by discarding any absolute reference frame and conceiving all motions as the relative motion of masses. In order to defend the concept of absolute space, Newton had considered a bucket filled with water and set into rotating motion, interpreting the inertial forces curving the water surface as an effect of absolute space. Mach raised the question of whether the same effect would also occur if the bucket were at rest and instead all the fixed stars were rotating around it. In this case, the effect

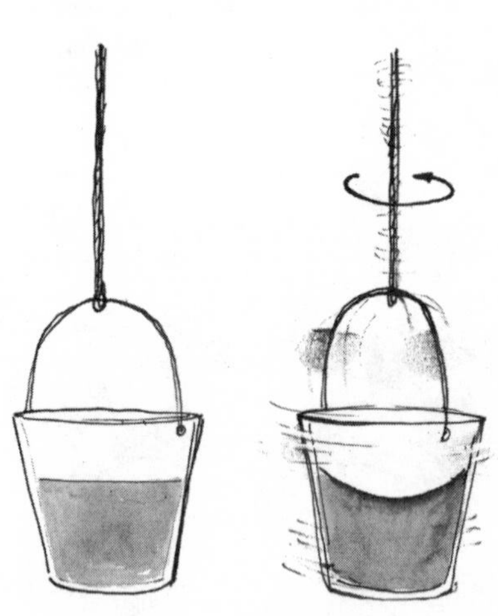

FIGURE 6.3. Newton's celebrated bucket experiment. Drawing by Laurent Taudin.

would be caused by the relative motion between the water and the stars, and the inertial forces curving the water surface would have to be ascribed to a distant interaction between masses, somehow comparable to gravitation.

Einstein's prior experience with electromagnetic field theory gave him a more precise mental model for conceptualizing the relation between the two forces. In developing special relativity as a response to the problems of the electrodynamics of moving bodies, he had realized that the presence of an electric or magnetic field depended on the frame of reference. Now he could similarly argue: "the gravitational field has a relative existence only in a manner similar to the electric field generated by magneto-electric induction."[12] In other words, with the help of Mach's interpretation of inertial effects as being due to an interaction between masses, Einstein developed the idea that gravitation and inertia form two aspects of one gravito-inertial field corresponding to the electromagnetic field.

Now the borderline problem of a relativistic field theory of gravitation had become much more specific. What was required was a field theory more or less analogous to electromagnetic field theory for the gravito-inertial field, with an appropriate field equation and an equation of motion. According to the field equation, matter tells the field how to behave, while according to the equation of motion, a field tells matter how to move.[13] Some of the implications of such a theory could be anticipated even before its formulation, in particular, the bending and redshift of light in a gravitational field and the slight rotation of Mercury's perihelion, which had been known since the nineteenth century. It was the only known deviation from Newton's theory that Einstein could possibly use to probe his new theory. Most deviation could be explained by disturbances by the other planets, with only 43 seconds of arc per century remaining unexplained. As early as 1907, Einstein expected that the new theory would explain this anomaly.[14]

To summarize: What motivated Einstein to launch the search for general relativity was an unexpected obstacle on his road to adapting all of physics to the new framework of special relativity. This obstacle brought him to question that very framework just two years after he had first established it. How could Einstein have known where to go with so few empirical clues? The answer in a nutshell is that he conceived the relation between gravitation and inertia as a borderline problem, transforming the available knowledge of classical physics in an astute way that was guided by the mental model of electromagnetic field theory.

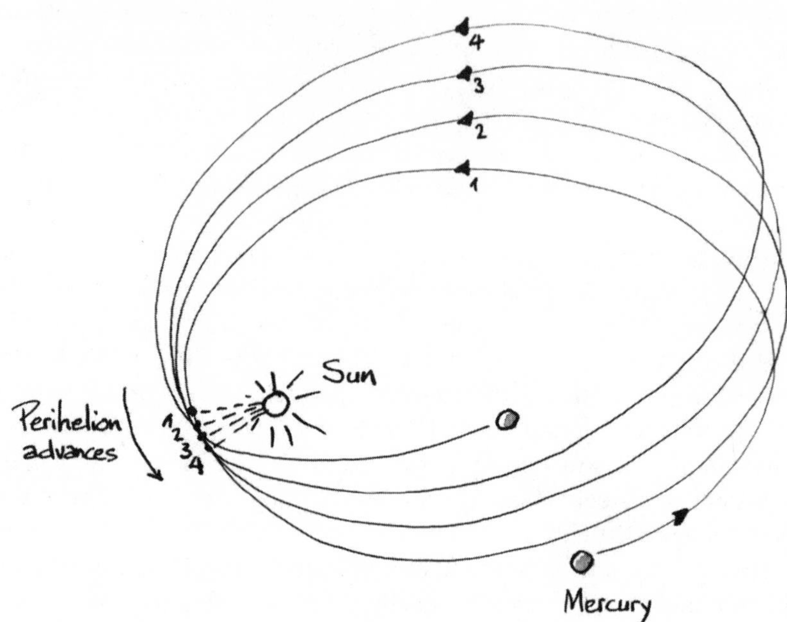

FIGURE 6.4. What causes the precession of the perihelion of Mercury's orbit? Drawing by Laurent Taudin.

The First Steps toward the Field Equations

At the beginning of 1911, Einstein became the chair of physics at the German University in Prague—his first full professorship. We have a later testimony from Otto Stern about Einstein in his office:

> I expected to meet a very learned scholar with a large beard, but found nobody of that kind. Instead, sitting behind a desk was a guy without a tie who looked like an Italian road mender. This was Einstein. He was terribly nice. In the afternoon he was wearing a suit and was shaven. I had hardly recognized him.[15]

From this time, a little notebook has been preserved that contains the first entries about the new theory of gravitation on which Einstein had begun working while he was in Prague.[16] In there, we find some simple expressions that are still a far cry from the complex later equations of general relativity, but that do have the same structure. In his search for a field equation, Einstein reformulated Newton's law of gravitation in field-theoretic terms and attempted to correlate it with the insights of the equivalence principle. In his search for an equation of motion, he could now simulate a dynamic gravitational field with the effects occurring in an accelerated frame of reference. He could consider, for instance, the inertial forces occurring in a rotating frame of reference, an example indicated by the discussions around Newton's and Mach's bucket experiments. According to Einstein's heuristics, it should have been possible to consider such a system at rest and the forces as being gravitational forces.[17]

This special case held a remarkable insight for Einstein. Considering a rotating disk in the context of special relativity, one could try to measure its circumference with little rods. Because of the length contraction predicted by special relativity, the rods will shrink so that one needs more of them to cover the entire circumference. In other words, the ratio between circumference and diameter will be larger than π, while the measurement rods along the radius will remain unaffected. This simple thought experiment suggested that one needs to go beyond Euclidean geometry to describe gravitational fields. Non-Euclidean geometries were well known at the time, but at that point Einstein wasn't too familiar with them.[18] He did realize, however, that what corresponds to a straight line in such a geometry is a geodesic line: a line of least curvature. Against this background, it soon became clear that a geodesic line would describe motion influenced by a gravito-inertial field when no other forces are present. A surprisingly simple answer was thus found to the question of the equation of motion which gave the entire problem a geometrical twist. Meanwhile, in 1907, the mathematician Hermann Minkowski had developed a geometrical formulation of special relativity in terms of a four-dimensional space-time continuum. This formulation became the basis for the more general mathematical framework required by general relativity.

This framework had far-reaching implications for the other, still open problem of the field equation. A reformulation of the equation of motion in terms of a geodesic line suggested that the gravitational potential of the new theory was given by the metric of a curved four-dimensional space-time, where the concept of "metric" generalizes the Euclidean concept of distance. In three-dimensional Euclidean space, the familiar metric instruction is to measure the distance between two points with the help of the Pythagorean theorem by forming the sum of the squares of their Cartesian coordinate separations. In Minkowski's space-time,

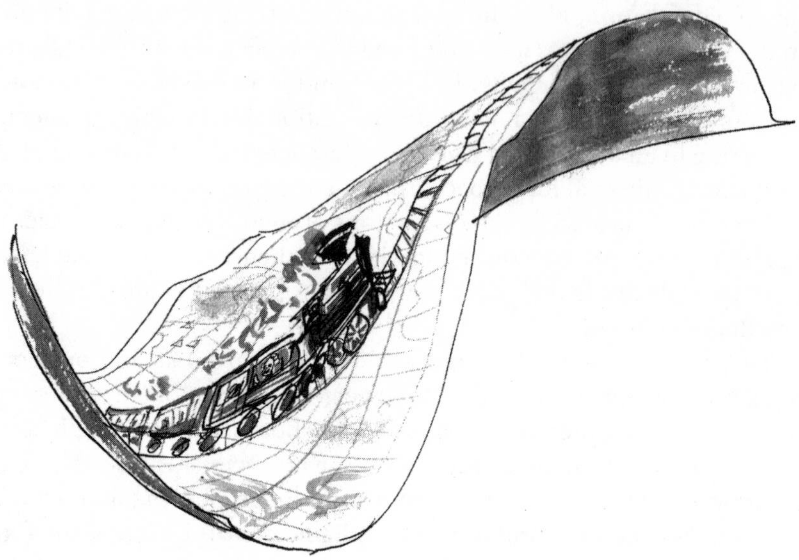

FIGURE 6.5. This is how a straight path looks on a curved surface. Drawing by Laurent Taudin.

the square of this distance is instead the square of the time separation between the two events, minus the square of their spatial separation. It is essentially an extension of the Pythagorean theorem to four dimensions, adapted to the special nature of the time coordinate. In a curved space-time, a variable metric associates different actual distances with a given coordinate distance at different locations on the surface. This variable metric turned out to be a suitable representation of the gravitational potential. The classical concept of gravitation as an attraction between masses was thus reconceptualized as a state of space-time, which is caused by and acts upon masses. Even more succinctly, in the words of John Archibald Wheeler: "Spacetime tells matter how to move; matter tells spacetime how to curve."[19]

While still in Prague, Einstein also achieved something else: for the first time, he related his theoretical efforts to astronomy. He contacted the young Potsdam astronomer Erwin Freundlich, the first astronomer to take his predictions seriously, and who later supported Einstein against the resistance of his superiors to verify the predictions of light bending and gravitational redshift.[20] In his first letter from Prague, Einstein wrote to him: "I would be delighted if you wished to tackle this interesting question. I know perfectly well that to answer it through experiment

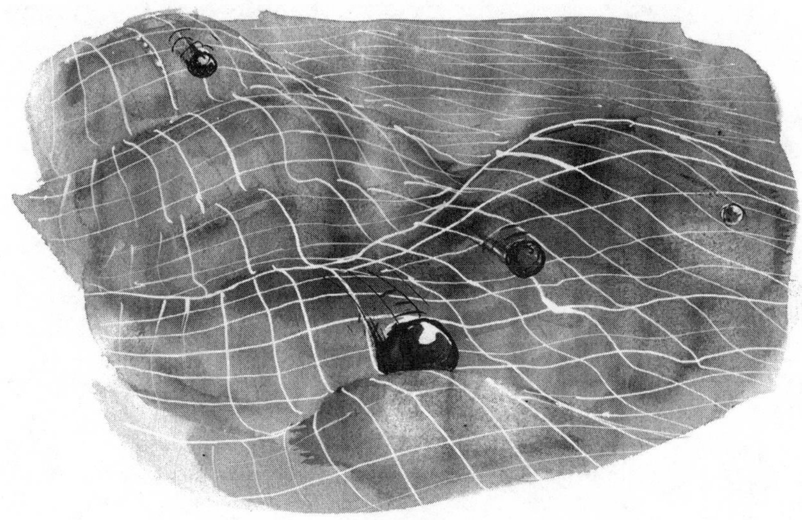

FIGURE 6.6. "Spacetime tells matter how to move; matter tells spacetime how to curve" (John Archibald Wheeler). Drawing by Laurent Taudin.

is no easy matter, for the refraction of the solar atmosphere may come into play."[21] On the occasion of a visit in Berlin, Einstein and Freundlich further explored potential astronomical consequences of the theory. In 1912, Einstein calculated the effect of gravitational lensing, the focusing of light by gravitation, a prediction he only published twenty-four years later (see figure 6.7).[22]

The Zurich Notebook—Einstein's Double Strategy

In the summer of 1912, Einstein returned to Zurich. He knew that he could describe a curved space-time using the metric tensor (the metric tensor replaces the one Newtonian gravitational potential with ten gravitational potentials). As we have seen, he was also able to write down the equation of motion in terms of the metric tensor: as the geodesic line in a curved geometry described by that metric tensor. But he had no clue how to find a field equation for this object. One of the most important sources for reconstructing Einstein's search for the field equation is a notebook—the so-called Zurich Notebook— in which he entered his calculations in the winter of 1912–1913.[23]

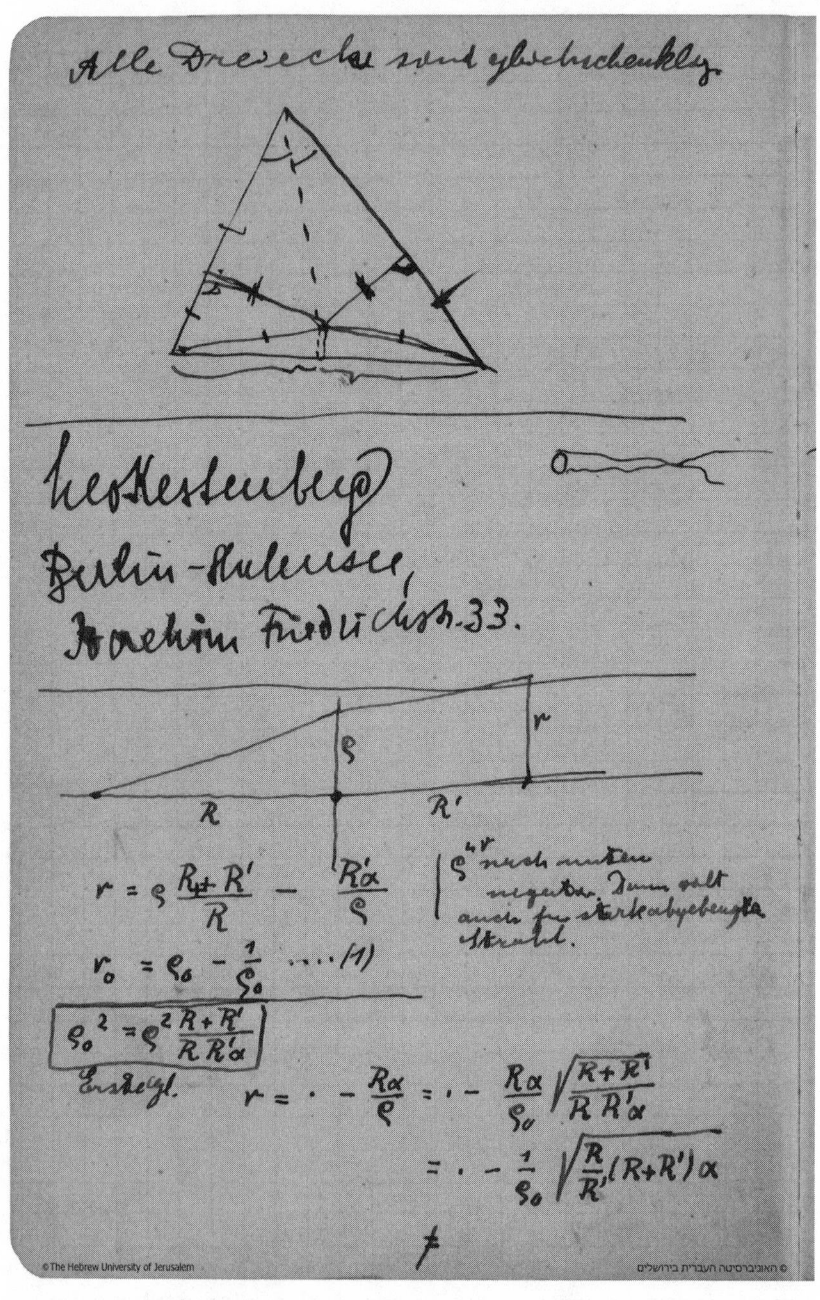

FIGURE 6.7. Notes about gravitational lensing dated to 1912 on a page of Einstein's Prague Notebook. Courtesy of the Einstein Archives, The Hebrew University of Jerusalem.

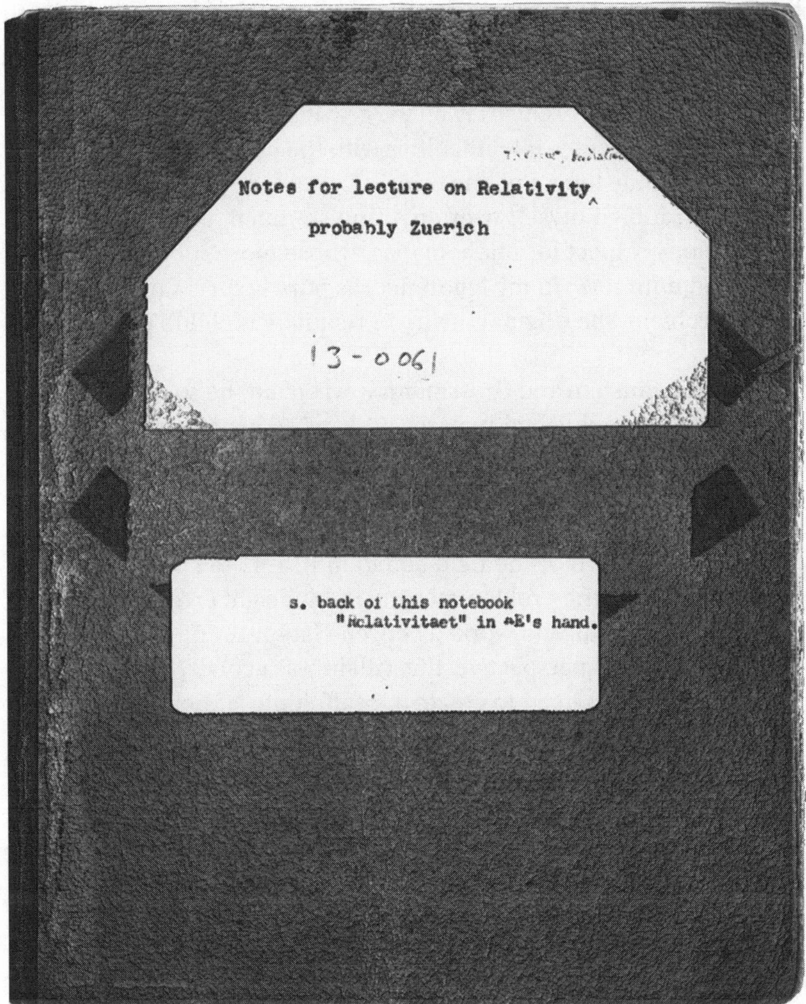

FIGURE 6.8. The Zurich Notebook, which contains Albert Einstein's research notes of 1912–13. Courtesy of the Albert Einstein Archives, The Hebrew University of Jerusalem.

The Zurich Notebook also documents Einstein's first, rather primitive attempts at dealing with the mathematics of the metric tensor. He attempted to bring the metric tensor together with what he knew at that point about the gravitational field equation, which was relatively little. It took Einstein three more years to solve this problem. His desperation with the mathematical challenges led him to call on his old friend Marcel Grossman for support. Grossman had helped Einstein earlier to survive his exams at the Zurich Polytechnic, and was instrumental in getting

Einstein a position at the patent office in Bern. Now he helped him to master the problem of gravitation. Einstein wrote to Arnold Sommerfeld:

> I'm now working exclusively on the gravitation problem, and believe that I can overcome all difficulties with the help of a mathematician friend of mine here. But one thing is certain, never before in my life have I troubled myself over anything so much, and I have gained enormous respect for mathematics, whose more subtle parts I considered until now, in my ignorance, as pure luxury! Compared with this problem, the original theory of relativity is child's play.[24]

Together, Einstein and Grossmann worked out the so-called *Entwurf* or draft theory, published in early 1913.[25] The historian John Stachel described the creation of general relativity as a drama in three acts, beginning with the equivalence principle in 1907, the realization that the gravitational potential is represented by the metric tensor in 1912, and the formulation of the field equation in 1915.[26] The villain of the story seems to be the problematic and in hindsight erroneous *Entwurf* theory on which Einstein worked between late 1912 and 1915. But as we shall see, from our perspective, this villain was actually the hero of the drama because it helped to create a "scaffolding" (Michel Janssen) on which the final theory could be built.[27]

Glancing at the notebook, one immediately recognizes Grossman's intervention. Einstein wrote his name next to the Riemann tensor, which was to become a crucial part of his new gravitational field theory. It turned out to be difficult to reconcile Einstein's physical expectations with the new formalism. These physical expectations comprised the requirements that the new theory should yield the familiar Newtonian law of gravitation under appropriate limiting circumstances and that it must comply with the conservation of energy and momentum. How could these expectations be reconciled with the ambition to generalize the relativity principle to accelerated motion? Exploring the new formalism during the winter of 1912–1913, Einstein and Grossman even hit upon the essentially correct field equation in a weak field limit, three years before the final paper was published.[28]

Einstein came to the conclusion, however, that this field equation did not meet his physical expectations. He eventually learned how to adapt and modify these expectations to the implications of the new formalism.[29] This learning experience is documented in the Zurich Notebook, as well as in contemporary publications and correspondence, and has been reconstructed in great detail by a team of researchers comprising

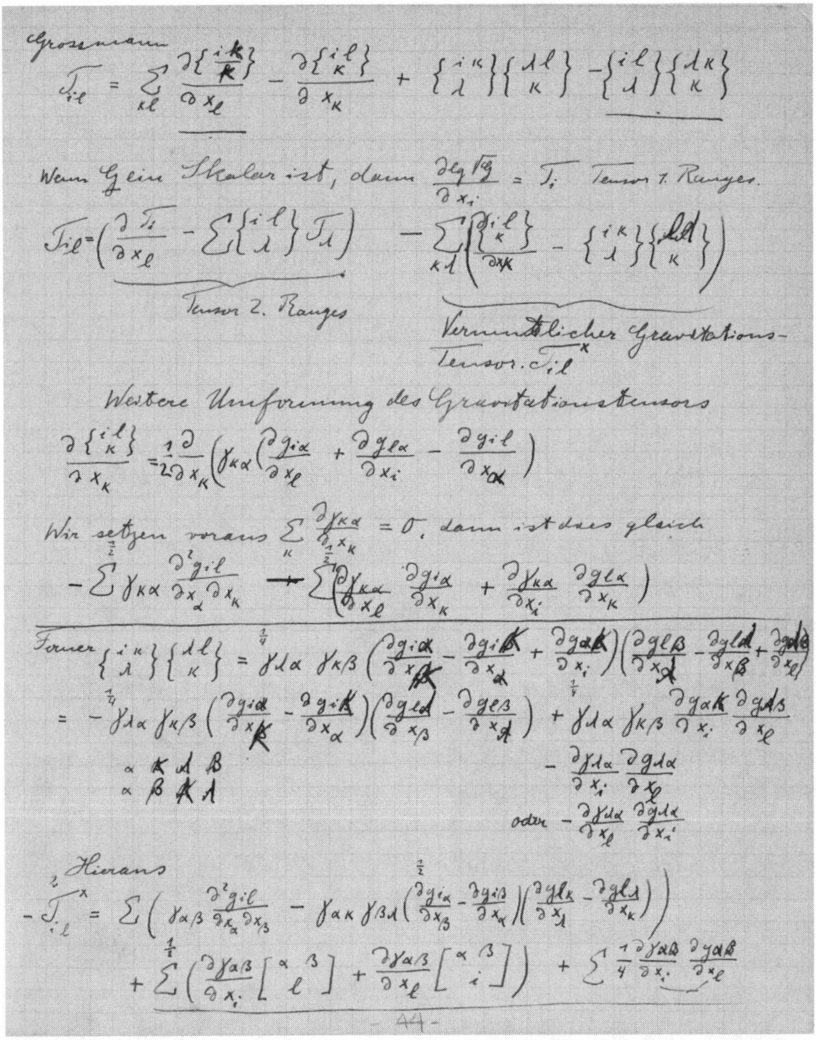

FIGURE 6.9. In the Zurich Notebook, Einstein began to look for field equations for his new theory of gravity with the help of Marcel Grossmann. At the top of this particular page, 22R, Einstein wrote down the generally covariant Ricci tensor T_{il} under the heading "Grossmann," presumably because Grossmann had suggested it to him. "If [the determinant of the metric] G is a scalar," Einstein noted, "the first half of T_{il} transforms as a tensor." Courtesy of the Albert Einstein Archives, The Hebrew University of Jerusalem.

Michel Janssen, John Norton, John Stachel, Tilman Sauer, and the author.[30] Einstein had started out by postulating a field equation with the help of a simple, "hand-made" mathematical formalism. The structure of the field equation was modeled on the basis of the classical field theory of electromagnetism: the source of the field—in this case mass and energy—generates the gravitational field.

Einstein's starting point had been a field equation delivering the Newtonian limit case as he expected it. But ideally, such a field equation, being "generally covariant," should hold in all coordinate systems, which suggested to him that it complies with his demand of a general relativity principle. Starting from a self-made formalism, this was hard to achieve. The Riemann tensor, on the other hand, to which Grossmann had introduced him, represents just such a generally covariant object. But how could one extract from it a field equation for which the Newtonian limit is correct and that is also in agreement with the principles of energy momentum conservation? The problem was that it was not entirely clear whether a field equation built on the Riemann tensor would actually deliver the requested physical results, and what kind of tweaking would be required to obtain them. In short, the mechanism was great, but the output uncertain.

FIGURE 6.10. Where to start? Mathematics or physics? Drawing by Laurent Taudin.

Given this situation, Einstein could pursue two different strategies. The physical strategy suggested he start with an object that fulfills the basic physical requirements and then check the extent to which it generalizes the relativity principle. The mathematical strategy suggested he begin with an expression that holds in all coordinate systems and then check whether the physical requirements are fulfilled. In the winter of 1912–1913, Einstein and Grossmann oscillated between these two strategies. In the end, they decided on the physical strategy, tweaking and adapting it as much as possible to also meet the requirements of a generalized principle of relativity.

The *Entwurf* Theory—A Crucial Intermediate Construct

The result of the collaboration between Einstein and Grossman was a hybrid theory, the *Entwurf* theory, mentioned earlier as the villain of the drama.[31] It was a hybrid theory because its equation of motion (telling matter how to move under the influence of a gravitational field) is generally covariant, retaining its form in all coordinate systems. In contrast, its field equation (determining the gravitational field) is not covariant. Einstein was nevertheless content with his achievement because it seemed the best he could reach. To his new love and future second wife Elsa he wrote:

> I finally solved the problem a few weeks ago. It is a bold extension of the theory of relativity, together with the theory of gravitation. Now I must permit myself some rest, otherwise I'll go kaput before long.[32]

But was this theory really worth the effort? Wasn't the *Entwurf* theory rather just a blind alley and a waste of more than two and a half years of his time? The question is important in understanding how theories change. Was Einstein's search for the gravitational field equation simply a comedy of errors, or can we understand it as an interesting intellectual development, a true learning experience?

This is how Einstein himself looked back on the genesis of general relativity:

> I have learned something else from the theory of gravitation: no collection of empirical facts, no matter how comprehensive can ever lead to the formulation of such complicated equations. . . . [they] can be found only through the discovery of a logically simple

> mathematical condition that determines the equations completely or almost completely. Once one has obtained those sufficiently strong formal conditions, one requires only little knowledge of the facts for the construction of the theory.[33]

This quotation from 1949 seems to confirm that the *Entwurf* theory, based on the physical strategy, which Einstein did not even bother to mention, had just been a blind alley, eventually giving way to the successful mathematical strategy.

Einstein's recollection was, however, colored by his later views (in particular by his struggle to find a unified field theory) and is therefore misleading, as the historian of science Jeroen van Dongen has argued.[34] The preponderance of contemporary evidence from the period 1913–1915 militates against the idea that it was a return to the mathematical strategy in the eleventh hour that paved the way to general relativity. Einstein himself warned:

> If you want to find out anything from the theoretical physicists about the methods they use, I advise you to stick closely to one principle: don't listen to their words, fix your attention on their deeds.[35]

What then was the exact function of the *Entwurf* theory for the creation of general relativity? It helped Einstein to assemble crucial insights on the way to the final theory, and to integrate them into a comprehensive mathematical framework that opened up new possibilities for his search. This framework eventually enabled a reconceptualization of some of the basic physical tenets that had formed his starting point. Due to its hybrid character, the *Entwurf* theory could serve as a bridge between physics and mathematics and also between Einstein's original heuristics and the new theory of general relativity. Many of the tools that later served to articulate and validate general relativity were first developed under its auspices.

In the summer of 1913, for instance, Einstein first calculated, with the help of his friend Michele Besso, Mercury's perihelion motion in the context of the *Entwurf* theory. Their calculations are documented in a manuscript and have been reconstructed in detail by Michel Janssen.[36] The value of the perihelion motion, which Einstein and Besso at first miscalculated, was too small. But this did not shatter Einstein's confidence in the theory. He never mentioned the result achieved with Besso's help, but later, in November 1915, reused the same techniques developed in the framework of the *Entwurf* theory to perform the correct calculation of Mercury's perihelion motion.[37] This calculation also played a crucial role in paving the way for the final theory, because it helped Einstein to revise his earlier understanding of the Newtonian limit, which had been a major stumbling block in his attempts to develop a generally covariant theory.

When Einstein redid this calculation in November 1915, now on the basis of the correct theory, his colleagues were amazed at how quickly he could calculate. The famous mathematician David Hilbert wrote him a postcard a day after Einstein had submitted the paper:

> Congratulations on conquering perihelion motion. If I could calculate as rapidly as you, in my equations the electron would correspondingly have to capitulate, and simultaneously the hydrogen atom would have to produce its note of apology about why it does not radiate.[38]

Actually, Einstein had only to redo the calculations for the perihelion motion in the *Entwurf* theory that he had done with Besso in 1913 but never published.[39] Einstein didn't bother to tell Hilbert about this earlier work, because—as Michel Janssen surmised—he apparently wanted to give Hilbert a dose of his own medicine, seeing Hilbert as someone who was "feigning the super-human through concealment of the methods."[40]

The *Entwurf* theory also acted as a stepping-stone toward the final theory in other respects. In 1914, with the help of another friend, Marcel Grossmann, Einstein developed a variational formalism from which the theory could be derived in a more elegant mathematical way.[41] It was also an attempt to bring the mathematical and physical strategies closer together. The development of a variational formalism (again in collaboration with Grossmann) helped him, in particular, to solve the other problem he had encountered with the mathematical strategy: the conservation of energy and momentum. With the new formalism, it turned out to be much easier to derive the conservation laws (actually anticipating Emmy Noether's famous theorems on the relation between invariances and conservation laws). This new mathematical tool turned out to have a larger horizon of possible applications than that for which it was originally intended, that is, for a new derivation of the *Entwurf* theory. A variational formalism may be described as a machine that can make other machines, in this case, generate field equations for a variety of different theories such as that of the *Entwurf* theory. It could be used, in particular, to generate a candidate field equation for the physical strategy, but it could also be employed for generating a candidate field equation for the mathematical strategy. In the fall of 1915, when Einstein became ever more skeptical about the *Entwurf* theory, he was able to repurpose the new variational formalism to construct other candidate field equations and, eventually, the final version of general relativity.[42]

One can learn something else from this story. The creation of general relativity in the crucial period between 1912 and 1915 was, to a surprising extent, a team effort involving close collaborators such as Grossman and Besso, but also competitors such Max Abraham, Gunnar Nordström, and David Hilbert.[43] Further contributions came from Paul Bernays, Adriaan Fokker, and Erwin Freundlich. In contrast, the prominent Berlin physicists in whom Einstein had set so much hope after his call to Berlin in 1914, were rather disinterested in his efforts. This is one of the ironies of Einstein's move to Berlin: he was expected to advance the new quantum theory with which he became ever more disenchanted, while focusing his efforts on general relativity—in the eyes of his established colleagues a hopeless endeavor. In early 1914, Einstein wrote to Besso:

> The fraternity of physicists behaves rather passively with respect to my gravitation paper. [. . .] Laue is not open to the fundamental considerations, and neither is Planck, while Sommerfeld is more likely to be so. A free, unprejudiced look is not at all characteristic of the (adult) Germans (blinders!). [44]

FIGURE 6.11. Einstein carried by "giants" (Newton, Maxwell, Gauss, Riemann) and "dwarfs" (Grossmann, Nordström, Freundlich, Besso) while Berlin's physics elite (Haber, Nernst, Rubens, Planck) looks on. Drawing by Laurent Taudin in the style of Albert Uderzo.

The Drama of November 1915—The Final Steps to the Field Equations

Einstein thus resolved the two major problems that had prevented him in Zurich from accepting candidate field equations found along the mathematical strategy, namely the requirement to comply with the Newtonian special case and the satisfaction of the conservation laws. But this raises a question: why did Einstein, once he had resolved these problems by the end of 1914, not return immediately to the mathematical strategy? It was a matter of perspective: he firmly believed in the *Entwurf* theory and had concocted all kinds of arguments in its favor, for instance, the infamous hole argument, which seemed to suggest that generally covariant theories cannot exist as a matter of principle.[45] But by October 1915, his perspective had gradually changed as problems with the *Entwurf* theory accumulated: As we have seen, it did not explain the perihelion problem; it also did not allow rotation to be conceived as being at rest, as Einstein suspected it should do because of his belief in a general principle of relativity in the spirit of Ernst Mach; and, as it turned out, it did not follow uniquely from a variational formalism. But this latter failure, as mentioned above, was a blessing in disguise. It meant that the formalism was actually more general and not just tailor-made for the *Entwurf* theory.

This was the situation that set the stage for the drama of November 1915 when Einstein published, week after week, his four conclusive publications on general relativity. On November 4, he published a new field equation based on the Riemann tensor (we call it the "November theory"),[46] with an addendum on November 11.[47] He then published the Mercury paper on November 18,[48] and the final field equations on November 25.[49] The first of these four papers constitutes the crucial step, switching from the *Entwurf* theory to a candidate field equation that complied with the expectations rooted in the mathematical strategy, that is, making contact with the mathematics of the Riemann tensor. The paper contains a crucial hint as to how Einstein accomplished this switch, passing from the *Entwurf* equation to the new "field equation." There he speaks of a "fatal prejudice" that had hindered him, and also indicates what the key to the solution was, as he formulated in a later letter to Arnold Sommerfeld.[50]

To understand what the fatal prejudice and what the key to the solution were, one has to take a closer look at the mechanism at work here, in particular at the way in which physical concepts relevant to the field equation are mathematically represented.[51] As we have seen, Einstein had realized in 1912 that the gravitational *potential* is represented by the metric tensor, but there was also the question of how to mathematically represent the gravitational *field*. The experience from classical physics suggested that a derivative operation was involved here, but in the new formalism there were different ways of implementing this expectation. Essentially, it was enough to redefine the gravitational field (changing a single element in the variational formalism developed for the *Entwurf* field equation) in order to arrive at the theory of November 4, 1915. Einstein thus characterized his earlier definition of the gravitational field as a "fatal prejudice" and his new definition as the "key to the solution." But as we have seen, this switch has to be seen in the wider context of the relation between the *Entwurf* theory and the new "November theory." The elaboration and exploration of the *Entwurf* theory made this transition and the identification of the fatal error possible in the first place. In turn, this transition enabled Einstein to carry much of the formalism and physical insights assembled under the aegis of the *Entwurf* theory over to the world of the Riemann tensor and to recontextualize some of his earlier findings.[52]

The successive papers straightened out the logical structure of the theory and, guided by the mathematical representation, adjusted some of the earlier physical conceptions such as the understanding of the Newtonian limit. This was actually a dramatic process for Einstein because he had to revise the architecture of his theory step by step. To Sommerfeld he wrote: "Unfortunately, I have immortalized the final errors in this struggle in the Academy."[53] To his friend Paul Ehrenfest, he ironically commented: "Einstein has it easy: every year he retracts what he wrote in the preceding year."[54] What contributed to the drama was that the mathematician David Hilbert was on his trail. Hilbert had presented his field equations to the academy in Göttingen on November 20, 1915, five days *before* Einstein.[55] He correctly used the Riemann curvature scalar in his variational formalism. He did not explicitly write down the field equations but could have easily calculated them.

But there are other, more important features that distinguish the two approaches: in the late 1990s, page proofs of Hilbert's paper, which itself was only published after some decisive revisions in early 1916, turned up.[56] These page proofs are dated December 6, 1915, and suggest that Einstein was not pipped at the post by Hilbert after all. They rather reveal that the original version of Hilbert's theory was conceptually closer to the *Entwurf* theory than Einstein's final version. Hilbert's theory of

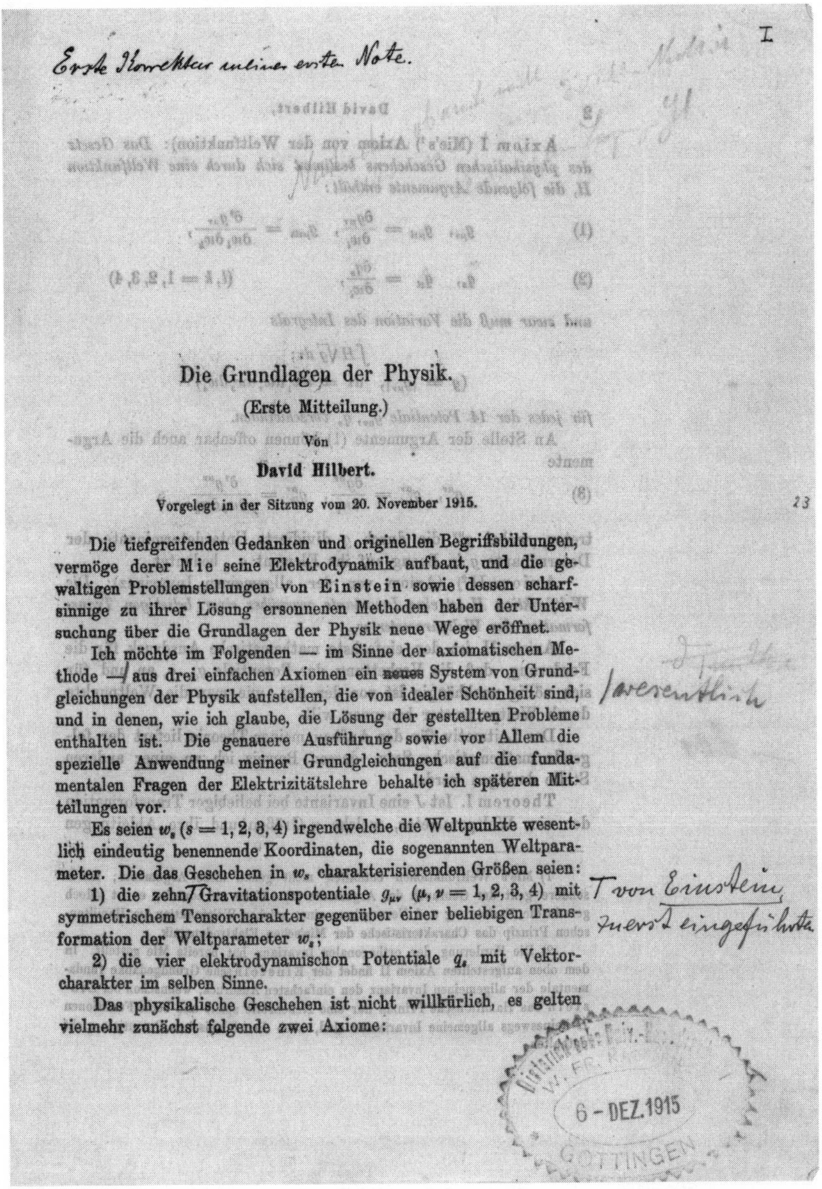

Erste Korrektur meiner ersten Note.

I

Die Grundlagen der Physik.

(Erste Mitteilung.)

Von

David Hilbert.

Vorgelegt in der Sitzung vom 20. November 1915.

Die tiefgreifenden Gedanken und originellen Begriffsbildungen, vermöge derer Mie seine Elektrodynamik aufbaut, und die gewaltigen Problemstellungen von Einstein sowie dessen scharfsinnige zu ihrer Lösung ersonnenen Methoden haben der Untersuchung über die Grundlagen der Physik neue Wege eröffnet.

Ich möchte im Folgenden — im Sinne der axiomatischen Methode — aus drei einfachen Axiomen ein neues System von Grundgleichungen der Physik aufstellen, die von idealer Schönheit sind, und in denen, wie ich glaube, die Lösung der gestellten Probleme enthalten ist. Die genauere Ausführung sowie vor Allem die spezielle Anwendung meiner Grundgleichungen auf die fundamentalen Fragen der Elektrizitätslehre behalte ich späteren Mitteilungen vor.

unwesentlich

Es seien w_s ($s = 1, 2, 3, 4$) irgendwelche die Weltpunkte wesentlich eindeutig benennende Koordinaten, die sogenannten Weltparameter. Die das Geschehen in w_s charakterisierenden Größen seien:

1) die zehn Gravitationspotentiale $g_{\mu\nu}$ ($\mu, \nu = 1, 2, 3, 4$) mit symmetrischem Tensorcharakter gegenüber einer beliebigen Transformation der Weltparameter w_s;

von Einstein zuerst eingeführte

2) die vier elektrodynamischen Potentiale q_s mit Vektorcharakter im selben Sinne.

Das physikalische Geschehen ist nicht willkürlich, es gelten vielmehr zunächst folgende zwei Axiome:

6 - DEZ. 1915

GÖTTINGEN

FIGURE 6.12. The first page of a set of proofs of Hilbert's first communication, with Hilbert's handwritten corrections and a printer's stamp, dated December 6, 1915. Reproduced with permission of Niedersachische Staats- und Universitätsbibliothek Göttingen Cod. Ms. D. Hilbert 634, Bl. 1.

December 6 was, just like the *Entwurf* theory, one that involved extra conditions on the admissible coordinate systems imposed by the requirement of energy momentum conservation. This restriction of coordinates was dropped in the published version. What still distinguished Hilbert's proposal from Einstein's was the generality of the source term. Whereas Einstein's theory admits all kinds of energy and matter, Hilbert had attempted to connect the theory of gravitation with a new theory of matter based on Mie's electrodynamic theory, adopting a special form of the source term of the gravitational field equation.[57] He thus created a research program that later Einstein himself would eagerly follow in his quest for a unified field theory. But in November 1915, Einstein could have taken his time and need not have worried about Hilbert stealing his thunder.

The Formative Years

It may thus seem as if, on November 25, 1915, the edifice of general relativity was complete. However, this was not the end of our story, and, in a sense, not even the beginning. While it is true that the equations formulated by Einstein have since stood the test of time, the elaboration of their consequences and their interpretation were just about to start and would have a profound impact on the understanding of the theory. General relativity had introduced a new language to describe nature that the scientific community was now attempting to master. Einstein himself had to learn it as well. It turned out, in particular, that the implications of the theory were at odds with his original motivations and ambitions.[58] Clearly, he had to abandon the "hole argument," for instance, which he had invented to support the *Entwurf* theory by arguing that a generally covariant theory would be unacceptable. In rethinking the argument with the help of the philosopher Moritz Schlick, Einstein quickly came to the conclusion that the flaw of the argument was the assignment of some physical reality to coordinate systems, while what actually mattered physically were only space-time coincidences.[59] This was a deep insight attained by rethinking and refuting his earlier heuristics.

Other elements of these heuristics were also difficult to reconcile with the insights gained by elaborating the newly established theory. The idea of extending the relativity principle from uniform to arbitrarily accelerated motions by imposing general covariance was, in 1917, seriously criticized by Erich Kretschmann.[60] Einstein's hope that general relativity

FIGURE 6.13. Albert Einstein and Willem de Sitter, Mount Wilson Observatory, January 1931. Reproduced with permission of Associated Press.

would realize Mach's vision that inertial effects can be explained as being due to an interaction of material bodies was challenged in the course of an extended exchange with the Dutch astronomer Willem de Sitter in the years between 1916 and 1930.[61] Einstein's related attempt in 1917 to establish a cosmological solution of general relativity describing a closed static universe (in which all inertia would be due to masses) was soon confronted by Alexander Friedmann and later by George Lemaître with alternative solutions describing an expanding universe.[62] Einstein was completely baffled and at first even incredulous with regard to the existence of such solutions.[63] But to his surprise, the idea of an expanding universe was confirmed in 1929 by observations of the redshift of distant galaxies at Mount Wilson observatory.[64] In short, within a few years, the entire conceptual framework, which had originally guided Einstein's pathway to general relativity, had been challenged.[65]

These were the *formative* years of general relativity (Hanoch Gutfreund and the author):[66] between 1915 and the early 1930s, a scientific community emerged around general relativity, the first exact solutions were found, beginning with the work of the astrophysicist Karl Schwarzschild, and first attempts were made at confirming predictions of the theory.[67] The confirmation of light bending during a British solar eclipse expedition led by Sir Arthur Eddington in 1919 turned Einstein overnight into an international star.[68] New insights were gained into the relation between the mathematical formalism and its physical interpretation, continuing the dynamics we observed in the Zurich Notebook. Relativistic cosmology emerged;[69] Einstein introduced the cosmological constant and then withdrew it.[70] As we have discussed, fundamental concepts underlying the theory were revisited and transformed. Attempts were made at moving beyond general relativity to create a unified theory of the gravitational and electromagnetic field or a quantum theory of gravity. The first fifteen years of general relativity constituted a vibrant period, which also saw the first discussions of gravitational waves. The formative years lasted roughly until the early 1930s, when interest in general relativity waned and more and more leading physicists turned their attention to the exploration of the booming field of quantum mechanics.

The Early History of Gravitational Waves

Given the prominent role of the analogy of gravitation with electromagnetism, the question of gravitational waves was a rather natural one. Several people had thought about it. One of the first was Max Abraham, who around 1912 was working on a gravitational field theory of his own.[71] He pointed out that gravitational waves differed from electromagnetic waves due to their lack of a dipole moment.[72] Einstein himself first explored the question of gravitational waves to some degree in the context of the *Entwurf* theory, that is, as early as 1913. In this year, Max Born asked Einstein how fast the effect of gravitation propagates according to the *Entwurf* theory. Einstein responded: "It is extraordinarily simple to write down the equations for the case where the disturbances one places into the field are infinitesimal."[73] Unfortunately, this is all we know about Einstein's thinking on this issue during the reign of the *Entwurf* theory.

After the completion of general relativity in 1915, the issue of gravitational waves demanded more serious attention. The first to bring it up was Schwarzschild.[74] In February 1916, he raised the question of the existence of gravitational waves according to the final theory in letters to Sommerfeld and Einstein. To Sommerfeld he wrote:

> I am rummaging around further in Einstein's field equations. Today I am totally flabbergasted. Setting up the problem of the plane gravitational wave according to Einstein one obtains the differential equation . . . [he writes down the equation and concludes] hence no wave motion but infinite speed of propagation.[75]

Evidently, he had found a clash between the qualitative idea of gravitation waves that seemed natural to him and his calculation. Einstein checked the calculation and responded: "Thus there are no gravitational waves analogous to light waves."[76]

Einstein put the problem to rest until another astronomer entered the scene. Willem De Sitter was, like Schwarzschild, one of the earliest members of the small community engaged in exploring the consequences of general relativity and beginning to challenge Einstein's own views on the subject. After Schwarzschild's letter, Einstein had come to believe that gravitational waves do not exist, pointing out that the approximation method used by Schwarzschild was not legitimate in the final theory. He wrote to Schwarzschild: "Since then, I have handled the Newtonian case differently, of course, according to the final theory."[77] But after what he had learned from De Sitter about approximate solutions, he became interested again. A letter from De Sitter spurred Einstein to return to the challenge of gravitational waves and helped him to overcome the hurdle found in Schwarzschild's treatment.[78] On June 22, 1916, Einstein submitted his first, famous paper on gravitational waves.[79]

The 1916 paper contains a calculation error, however, which even affected the qualitative assessment of gravitational waves, erroneously allowing monopole radiation to be emitted. Einstein uncovered and eventually corrected his error in the context of the dialogue with another colleague, the Finnish physicist Gunnar Nordström, who had taken issue with Einstein's treatment of energy-momentum conservation in general relativity.[80] In 1918, reacting to a detailed criticism by Nordström, Einstein published a revised paper on gravitational waves in which he stated:

> The important question of how gravitational fields propagate was treated by me in an academy paper one and a half years ago. However, I have to return to the subject matter since my former presentation is not sufficiently transparent and, furthermore, is marred by a regrettable error in calculation.[81]

It was this paper that became the reference point for the further investigation of gravitational waves, presenting the essentially correct form of the famous quadrupole formula.

This was not Einstein's last word on the issue. In 1936, he wrote a joint paper with his collaborator Nathan Rosen in which they claimed that gravitational waves do not exist as exact solutions of the theory.[82] They submitted their manuscript to the *Physical Review* where it was examined by a referee (the American mathematician and cosmologist Howard P. Robertson), who found a flaw in their argumentation.[83] Unfamiliar with the refereeing process, Einstein protested to the editor John T. Tate:

> We (Mr. Rosen and I) had sent you our manuscript for *publication* and have not authorized you to show it to experts before it is printed. I see no reason to discuss the—by the way erroneous— explanations of your anonymous referee.[84]

Einstein and Rosen withdrew their paper and published it in a different journal with a different conclusion, now showing that rigorous solutions for cylindrical gravitational waves exist.[85] But even after this breakthrough, Einstein's closest collaborators, Rosen and Leopold Infeld, continued to argue against the physical existence of gravitational waves, perhaps also under Einstein's influence.[86]

The Low-Water Mark Period

It would be shortsighted to consider this episode as simply a comedy of errors or an indication of Einstein and his collaborators having lost

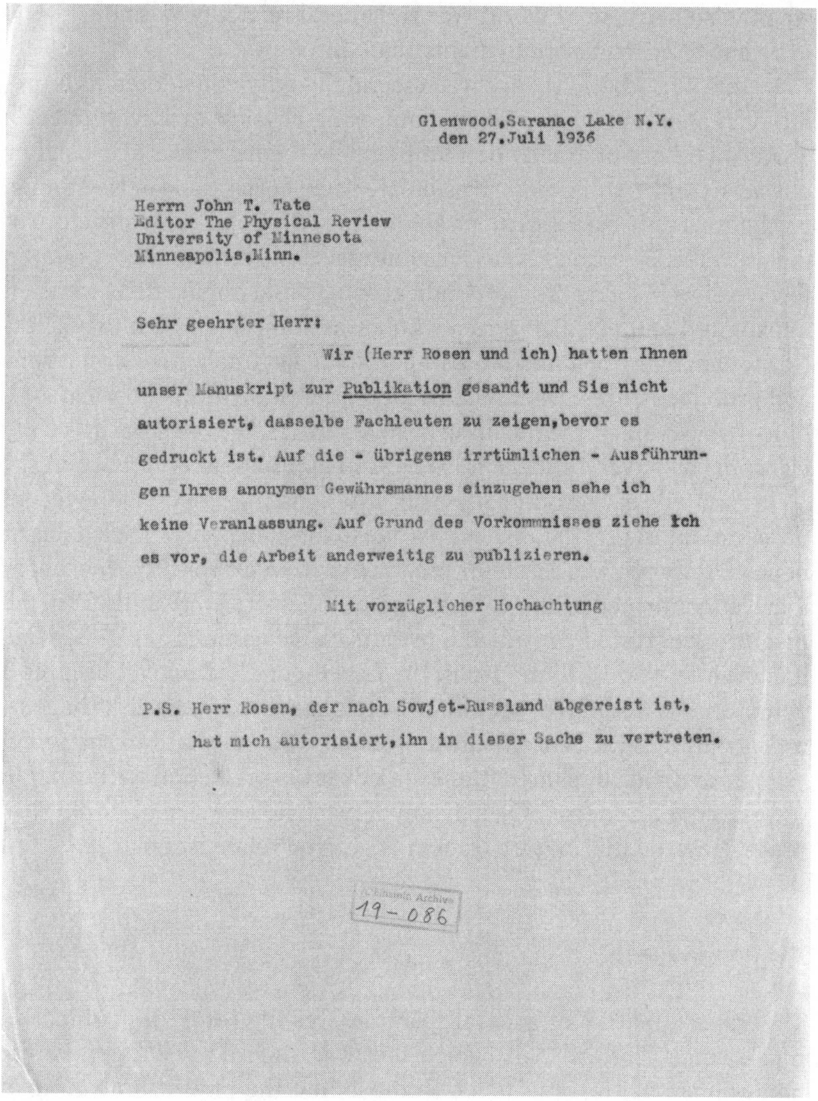

Glenwood, Saranac Lake N.Y.
den 27. Juli 1936

Herrn John T. Tate
Editor The Physical Review
University of Minnesota
Minneapolis, Minn.

Sehr geehrter Herr:

Wir (Herr Rosen und ich) hatten Ihnen unser Manuskript zur Publikation gesandt und Sie nicht autorisiert, dasselbe Fachleuten zu zeigen, bevor es gedruckt ist. Auf die - übrigens irrtümlichen - Ausführungen Ihres anonymen Gewährsmannes einzugehen sehe ich keine Veranlassung. Auf Grund des Vorkommnisses ziehe ich es vor, die Arbeit anderweitig zu publizieren.

Mit vorzüglicher Hochachtung

P.S. Herr Rosen, der nach Sowjet-Russland abgereist ist, hat mich autorisiert, ihn in dieser Sache zu vertreten.

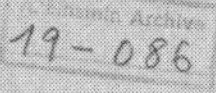

FIGURE 6.14. Einstein's letter to John T. Tate, the editor of the *Physical Review.* Courtesy of the Albert Einstein Archives, The Hebrew University of Jerusalem.

touch with contemporary physics. The reasons for the lack of clarity about the existence of gravitational waves are more profound and reflect the general state of general relativity until the mid-1950s. In fact, its formative years were followed by a low-water mark period (Jean Eisenstaedt), beginning in the early 1930s and ending during the 1950s when, also because of World War II, general relativity was considered to be basically irrelevant to mainstream physics.[87]

In this period, the theory was essentially only considered as a preliminary step toward a more encompassing physical theory, whether a quantum theory of gravity or a unified field theory. It was also used by physicists and astronomers to calculate small corrections of Newtonian predictions or to find kinematic solutions to the large-scale structure of the universe. Beyond a few exact solutions, however, the theory offered few physical insights. Its more radical conceptual implications were left unexplored and the theory as a whole was subjected to a more or less Newtonian interpretation. More important, it left fundamental questions and problems unresolved, such as the initial value problem, the definition of observables, the meaning of singularities, the question of energy transport by waves, or the admissibility of approximation procedures.[88]

Against this background, it makes little sense to claim that the recent observations of gravitational waves due to the collision of black holes or neutron stars confirm Einstein's predictions from 1916 or 1918. The theory that was confirmed by these observations differed significantly from the patchwork of results for which Einstein's field equations of 1915 had laid the groundwork. Until the 1950s, the new theory had not yet become a comprehensive, shared, and intuitively plausible framework that was universally applicable to physical problems. But only within such a developed and socially shared framework does the prediction of gravitation waves and the concept of a collision of black holes make clear physical sense. How did this mature version of general relativity emerge?[89]

The Renaissance

The transformation of general relativity was the result of its so-called renaissance (Clifford Will) which began in the mid-1950s, generating new conceptual insights (such as those into the physical meaning of gravitational waves or into the nature of black holes) and witnessing new astrophysical discoveries (such as that of the quasars in the 1960s).[90] During this renaissance, the theory underwent a profound epistemic transformation that resulted in a framework in which the physical objects considered could now for the first time be clearly defined. The same holds true for the criteria to be applied to the observational and experimental validation of predictions, such as that of gravitational waves. The new dynamics eventually developed into a veritable "Golden Age"

(Kip Thorne), which consolidated and elaborated the novel conceptual insights, turned the theory into the foundation of modern astrophysics and observational cosmology, and was crowned by the discovery of gravitational waves.[91]

The question of how this epistemic transformation of general relativity into a universally applicable physical framework came about touches a key issue of the history of science: the emergence of conceptual novelties and of theory transformation. These novelties clearly were not just the result of observational breakthroughs (such as the discovery of quasars or the observation of gravitational waves) because these breakthroughs were actually *preceded* by the relevant theoretical advances. Neither were these advances simply a consequence of the mathematical elaboration of Einstein's 1915 field equations because the epistemic transformation of general relativity involved an entirely new perspective, also concerning its mathematical framework.

Until the mid-1950s, research on general relativity was concentrated in a few international centers. However, the physicists and mathematicians involved did not yet form a coherent scientific community but pursued questions that were splintered into various research programs with little relation among each other. These research programs ranged from attempts to create a unified field theory in the tradition of Hilbert and Einstein, via steady-state cosmology, to attempts at creating a quantum theory of gravity. But even when scientists were working on similar problems, they were often not even aware of it. What these different perspectives had in common was the conviction that general relativity was little more than a springboard for a more comprehensive physical theory.

During the 1950s, this situation changed.[92] Relations between the seemingly isolated problems became evident and gradually a community of scientists working on these problems emerged. This change was due, on the one hand, to a general increase in the international exchange of scientists, fostered in particular by newly created postdoctoral programs.[93] It was fostered, on the other hand, by an international conference held in 1955 in Bern, on the occasion of the fiftieth anniversary of special relativity.[94] The meeting happened to coincide with the death of Albert Einstein, and the physicists assembled in Bern now somehow felt responsible for his legacy.[95] The Bern meeting did not immediately lead to a convergence of the different research agendas, but it did highlight the deficits in the state of knowledge on general relativity and, in turn, also increased the postdoctoral exchange on topics related to general relativity in the following years, thus giving rise to a self-sustaining process.[96]

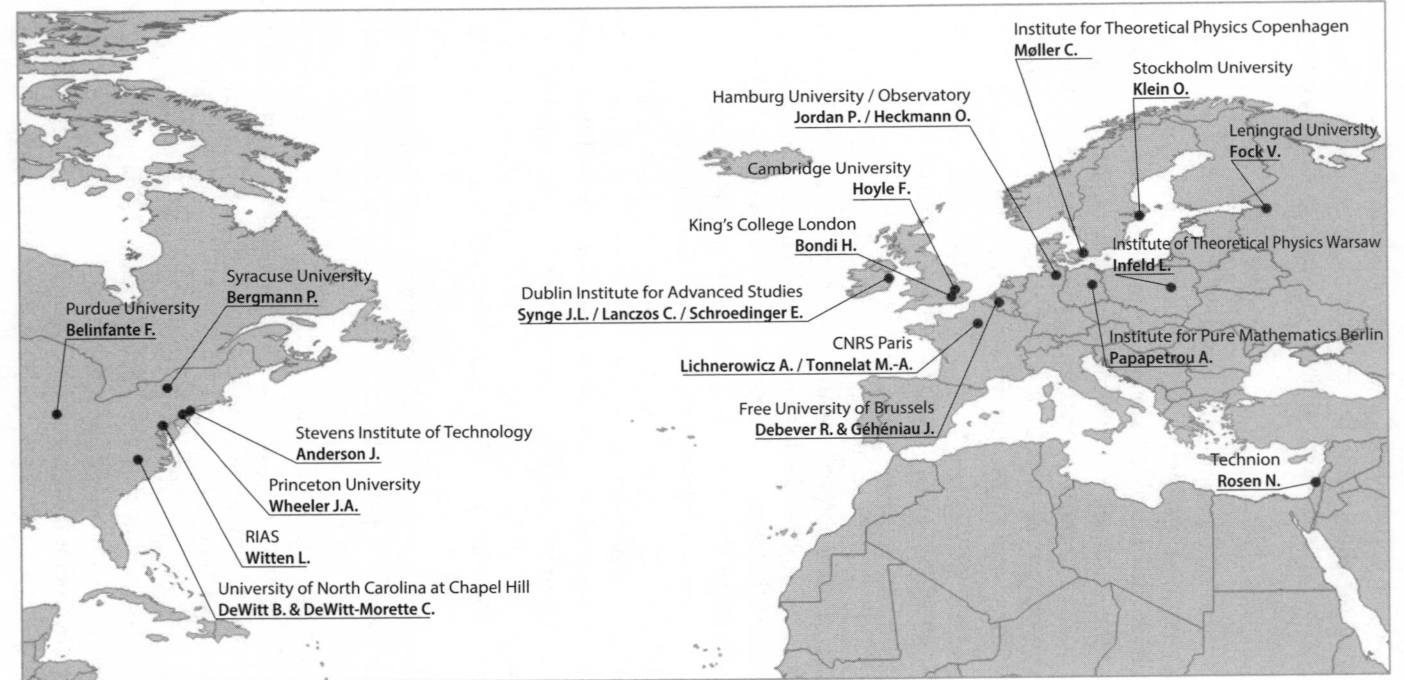

FIGURE 6.15. Research centers working on general relativity in 1955. The figure displays research centers based in North America and Europe that were working on different research agendas related to general relativity in 1955. With research centers we mean those institutional frameworks with at least one principal investigator who had an institutional position stable enough to attract postdocs and/or produce new PhDs in the field. Under the names of the institutions, one finds the names of the leading scientists in general relativity–related research projects in 1955. Image by Florian Kräutli and Roberto Lalli.

Long before many of the open problems of general relativity could actually be resolved, the insight into their interrelation led to a substantial change in their interpretation and meaning. It became clear, in particular, that general relativity could hardly offer a generally applicable physical framework as long as physical observables could not be defined and as long as the initial value problem was not precisely formulated. These theoretical questions were now generally acknowledged as relevant research issues. As a consequence, problems such as the nature of the gravitational degrees of freedom and the physical meaning of gravitational waves were systematically explored. The new awareness also transpires from introductions to contemporary publications, pointing to the difficulties in extracting unambiguous physical assertions from the theory.

FIGURE 6.16. John Lighton Synge's lecture at the international conference held in Warsaw and Jablonna in July 1962. This conference was the fourth international conference dedicated to topics related to general relativity. It was later referred to as GR3, whereas relativists christened the Berne conference "GR0" to imply that the conference was a starting point for the stable tradition of international conferences, a tradition that continues to this day. In the first row, left to right: Leopold Infeld, Vladimir Fock, James L. Anderson, Ezra Ted Newman, Roger Penrose, Banesh Hoffmann. At the far right: Roza Michalska-Trautman. Photograph by Marek Holzman, as published in *Relativistic Theories of Gravitation* (Oxford: Pergamon Press, 1964).

In the sequel to efforts responding to these challenges, new concepts and mental models emerged. They ranged from a precise characterization and classification of exact solutions, via a clearly defined physical model explaining how gravitational waves transfer energy to material detectors, to a reformulation of Einstein's field equations, bringing them into a form that renounced explicit general covariance but was more appropriate for unambiguous numerical calculations and computer simulations. The ultimate success story of gravitational waves and the international support for projects dedicated to their investigation can thus be seen as a result also of this epistemic transformation of general relativity (Alexander S. Blum, Roberto Lalli, and the author).

Without answers to the question of the role of boundary conditions, of the meaning of singularities, and of the relation between exact and approximate solutions, the further investigation of gravitational waves would hardly have been possible. The answers strengthened the trust in approximation schemes; they opened up paradigmatic ways for the application of the theory; and they provided the basis for dealing with specific issues such as calculation methods, the modeling of sources, and the design of detectors—without having to continually address the foundational issues of physics.

Beginnings of the Search for Gravitational Waves

The experimental search for gravitational waves was rooted in the same development described above.[97] It started immediately after the basic conceptual issues had been settled and was pioneered by Joseph Weber in the late 1950s. He constructed separate resonance detectors consisting of large aluminum cylinders for coincidence measurements. Weber was an electrical engineer who had contributed to the development of the MASER. His PhD supervisor, John Archibald Wheeler, was one of the protagonists of the renaissance of general relativity. The first ideas for a detector using an interferometer also go back to the early 1960s, to a Russian team whose project was, however, not pursued.

In 1969, Weber claimed to have discovered gravitational waves.[98] But by the mid-1970s, this claim was widely considered to be unsubstantiated. Meanwhile, however, Weber's claim had spurred research activities dedicated to gravitational waves all over the world. They eventually triggered the foundation of several international large-scale research projects, which included not only the building of large detectors but also

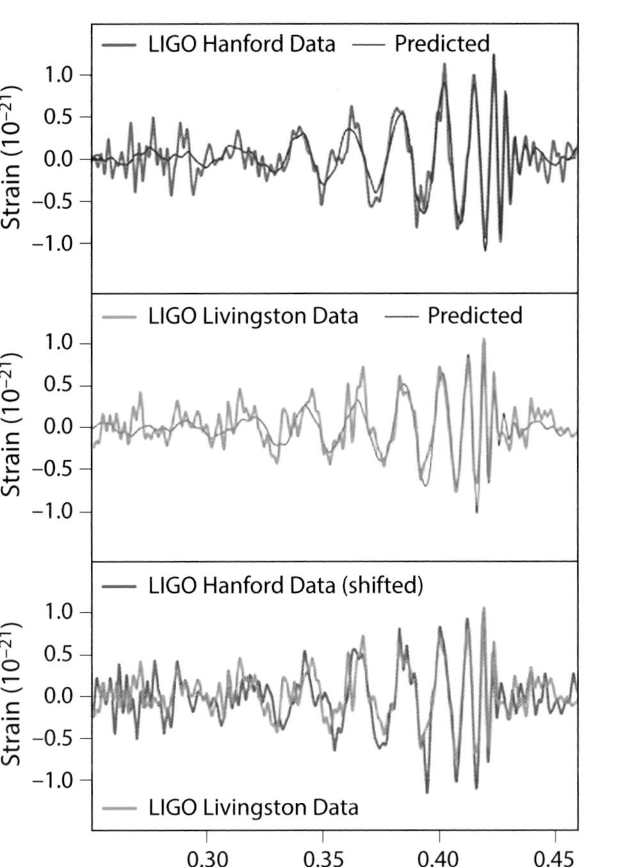

PLATE 1. The discovery figure.

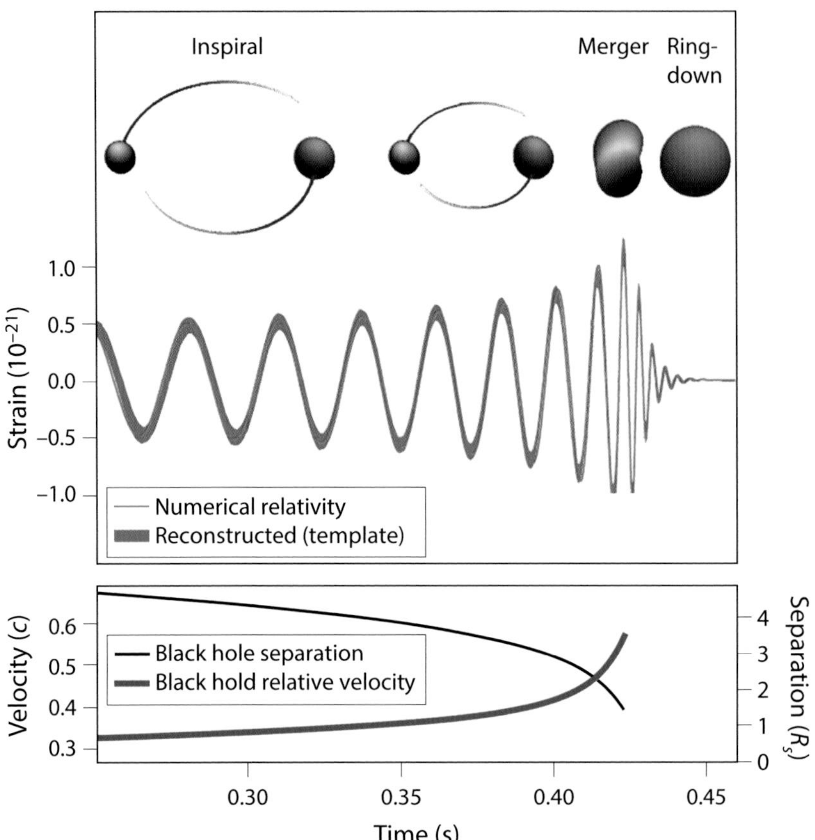

PLATE 2. Features of a compact binary merger.

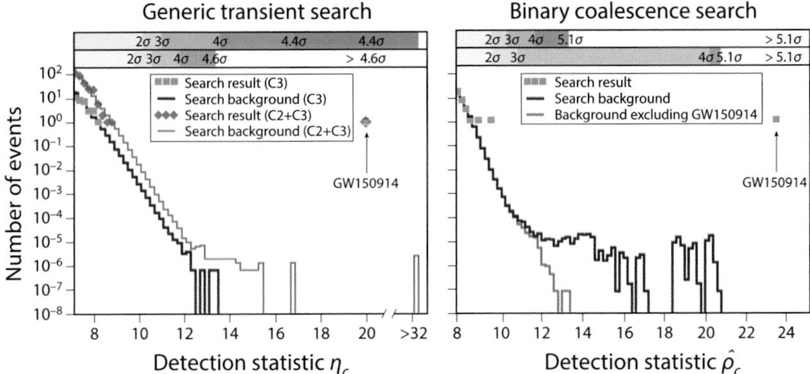

PLATE 3. Coincidence slices for the two detectors.

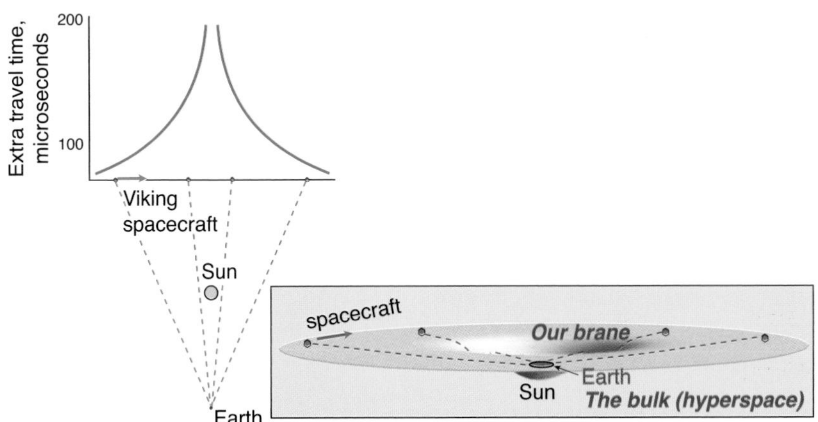

PLATE 4. Reasenberg-Shapiro Experiment that measured the warping of space around the sun using round-trip radio signals between Earth and the *Viking* spacecraft.

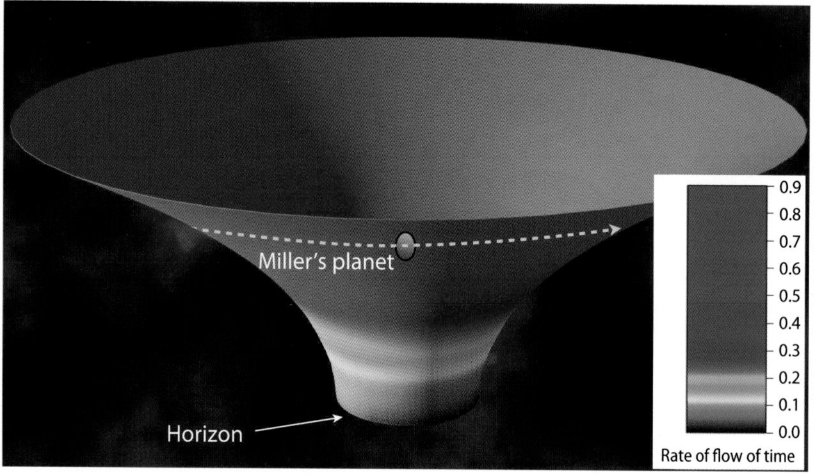

PLATE 5. Equatorial slice through a non-spinning black hole, as reviewed from the bulk, together with the smallest orbit hat Miller's planet can have and not plunge into the hole.

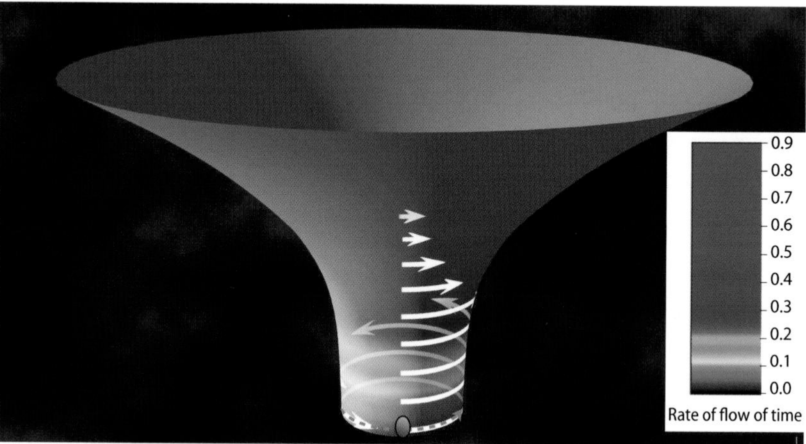

PLATE 6. The warped space of the black hole Gargantua, which spins very rapidly. The white arrows depict the hole's dragging of space into a whirling motion, which stabilizes the orbit of Miller's planet, permitting it to be very close to the hole's horizon.

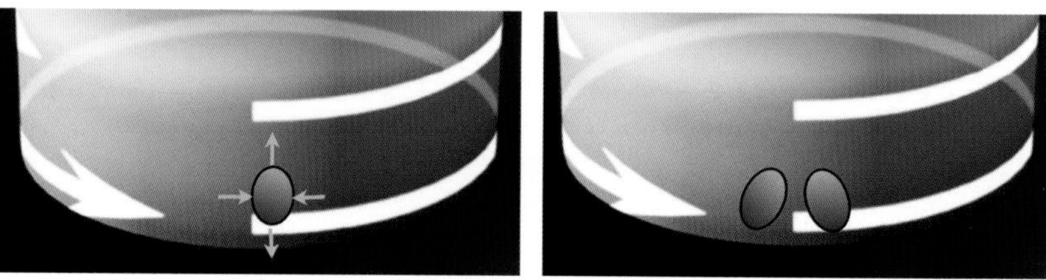

PLATE 7. *Left*: Miller's planet is deformed by Gargantua's tidal gravity. *Right*: The deformed planet rocks back and forth, producing a sloshing of the planet's oceans that results in giant water waves.

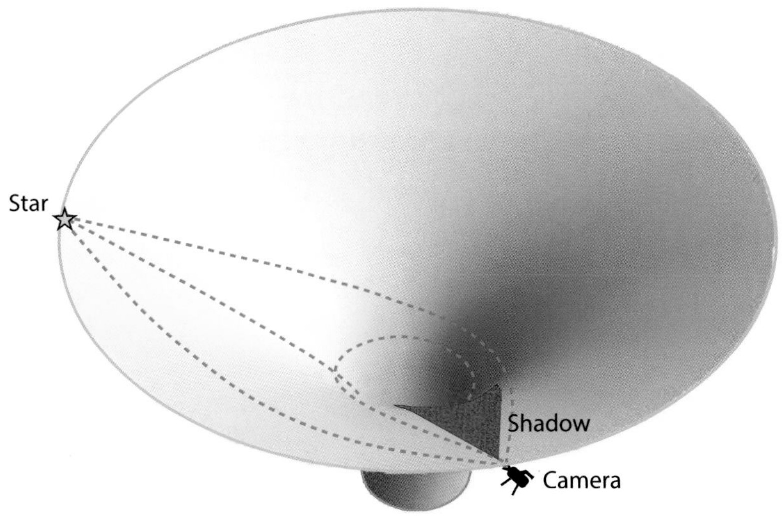

PLATE 8. Three light rays that travel along three different paths through the warped space of a black hole, from a star to a camera.

PLATE 9. A field of many stars gravitationally lensed by a fast-spinning black hole.

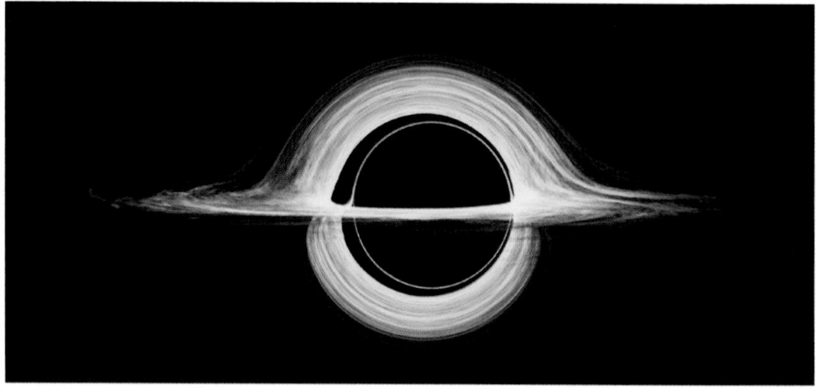

PLATE 10. A variant of the image of the black hole Gargantua with a thin disk of hot gas in its equatorial plane, as seen in *Interstellar*.

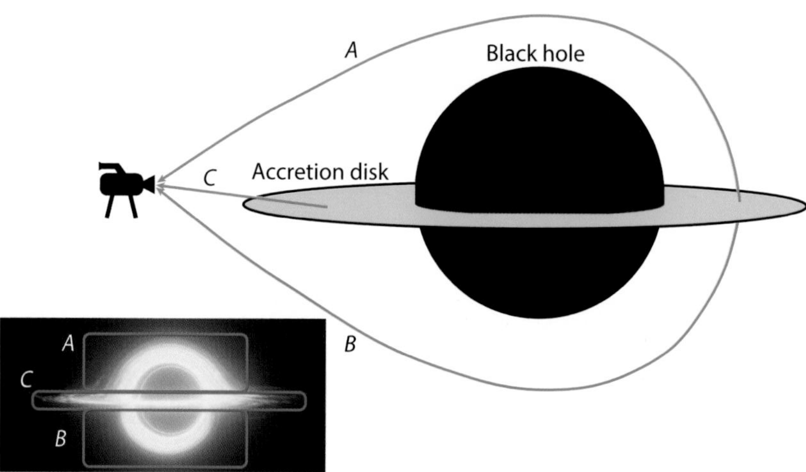

PLATE 11. The iconic image of the black hole Gargantua is produced by gravitational lensing of light from its thin, equatorial accretion disk.

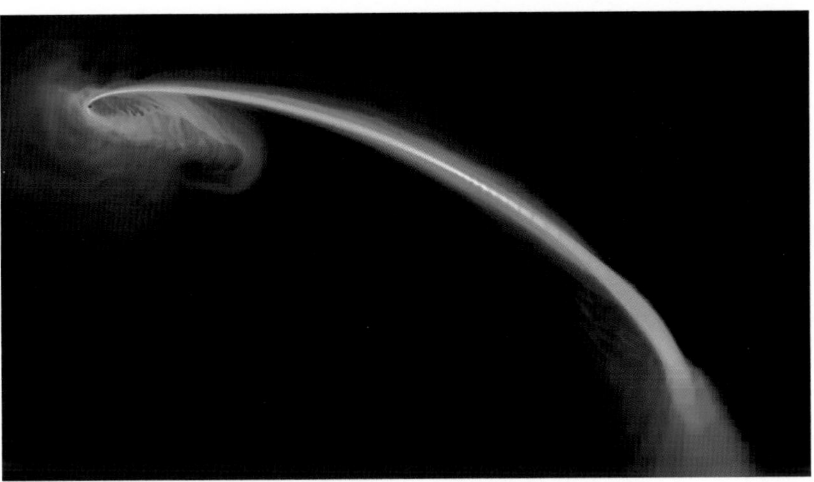

PLATE 12. Hot gas from tidal disruption of a star by a black hole.

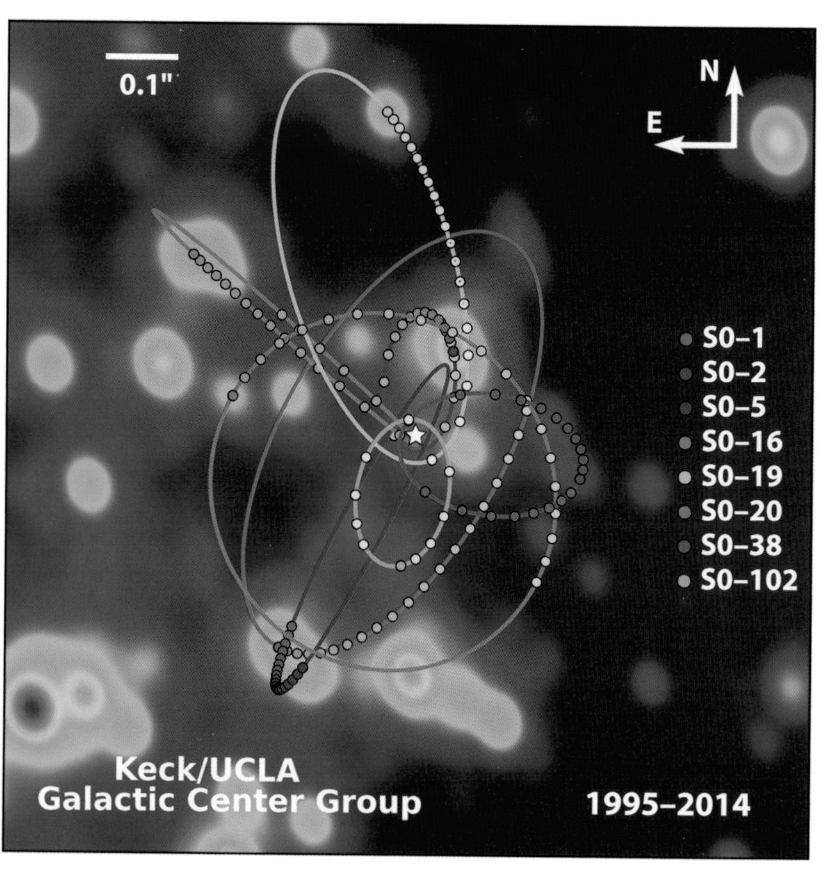

PLATE 13. Orbits of stars around the black hole at the center of our Milky Way galaxy.

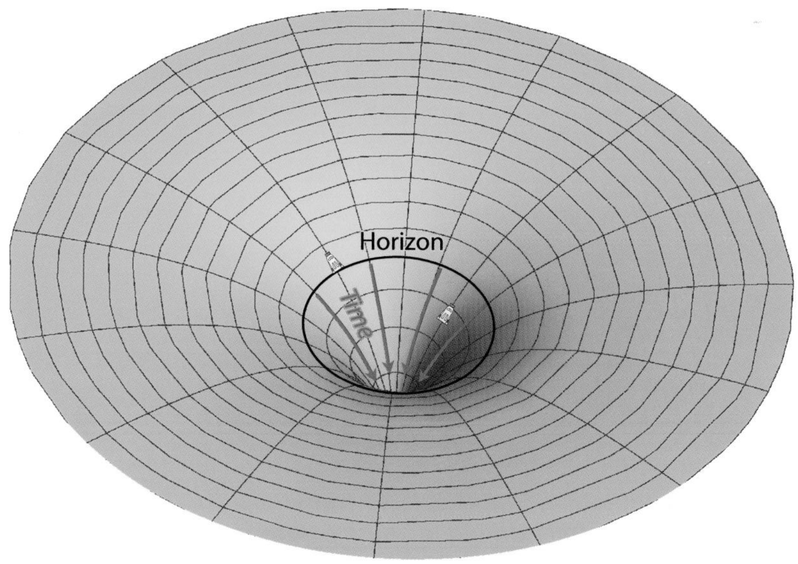

PLATE 14. A black hole as seen from the bulk. Time slows to a halt in a spacecraft hovering at the horizon.

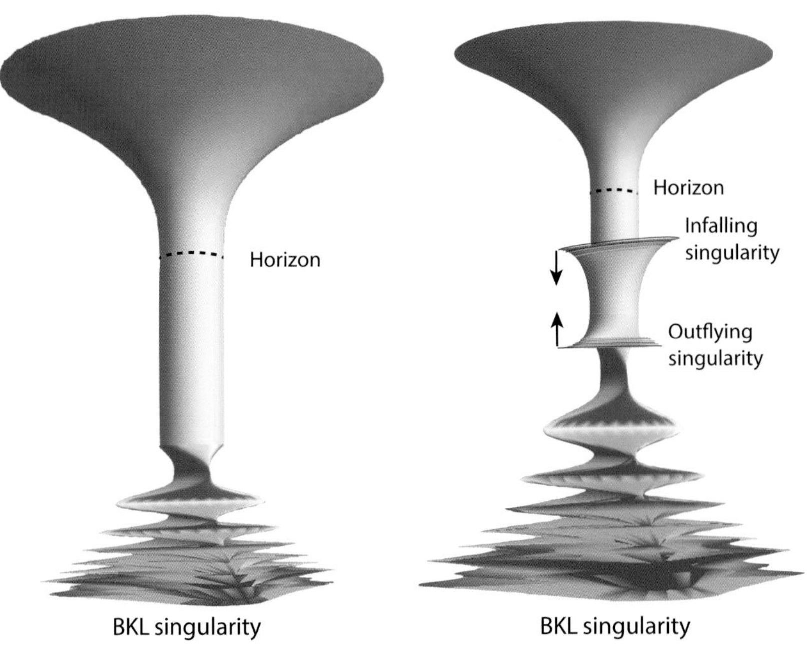

PLATE 15. *Left*: Heuristic drawing of the chaotic BKL singularity inside a black hole, as seen from the bulk. *Right*: The infalling and outflying singularities produced by stuff that falls into the black hole.

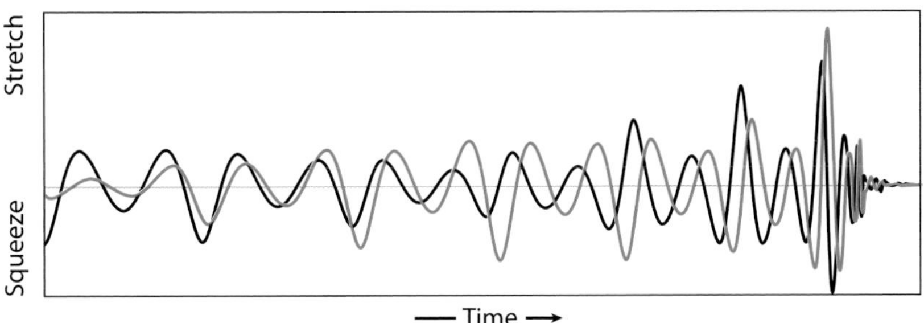

PLATE 16. Two gravitational waveforms, one red and one black, produced by black holes that orbit each other and then collide.

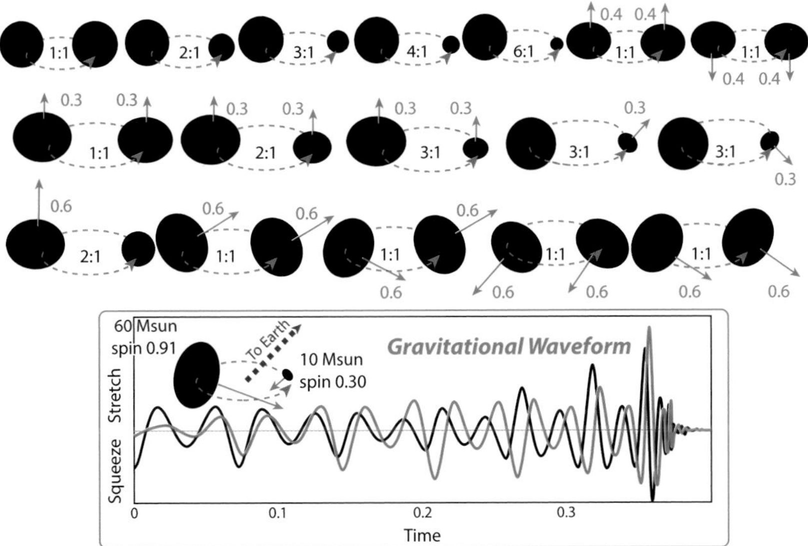

PLATE 17. *Top*: Seventeen of the black hole binaries whose waveforms are in the SXS catalog; 4:1 means a mass ratio of 4 to 1, the red arrows indicate the directions and magnitudes of the black holes' spins, the red numbers are the spin magnitudes as a fraction of the maximum possible spin, and the green arrows depict the orbit. *Bottom*: The two waveforms of a specific example of the binary's parameters. (A gravitational wave has two waveforms, just as light has two polarizations; LIGO measures only one of them.)

PLATE 18. The black hole binary that produced LIGO's first gravitational-wave signal, GW150914, as your eyes would have seen it if you had been nearby.

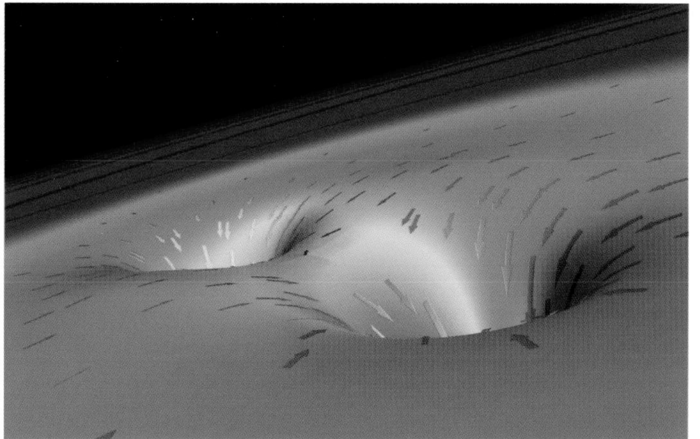

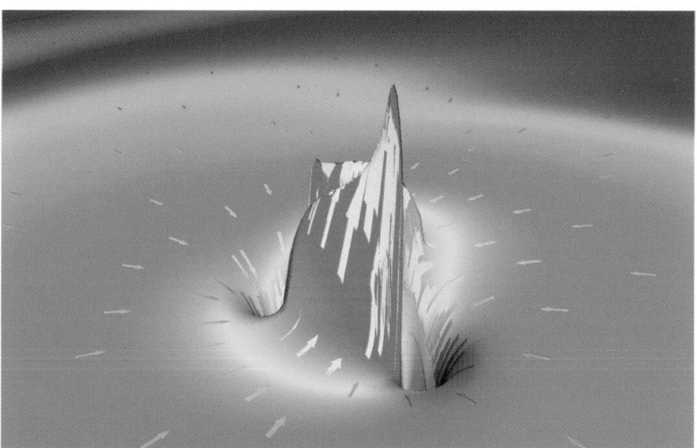

PLATE 19. Snapshots from a movie depicting the geometry of spacetime around the GW150914 binary black hole, as seen from the bulk, 60 ms before collision, at the moment of collision, and 12 ms after the collision.

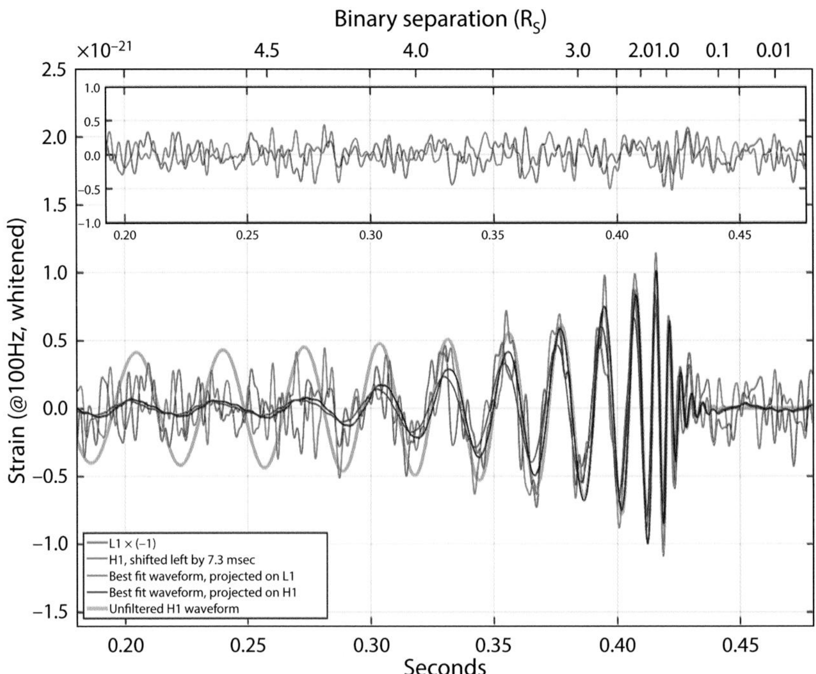

PLATE 20. The gravitational-wave event GW150914 observed by Advanced LIGO's Livingston, L1 (green) and Hanford, H1 (red) detectors, also showing best-fit templates computed by combining analytical and numerical activity. Data for L1 are shifted by 7.3 ms to account for the time of travel between detectors.

PLATE 21. Snapshots of numerical-relativity simulation of the binary black-hole coalescence of the gravitational-wave event GW151226.

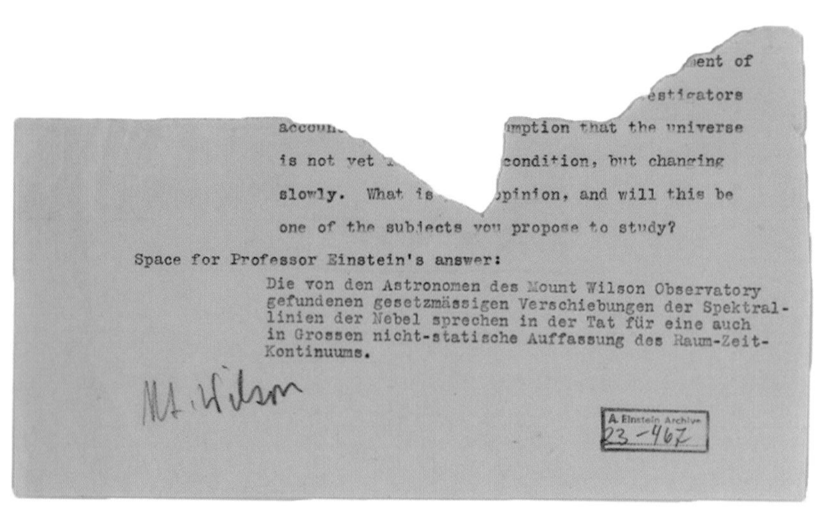

ment of
estigators
accoun... ...mption that the universe
is not yet ... condition, but changing
slowly. What is ... opinion, and will this be
one of the subjects you propose to study?

Space for Professor Einstein's answer:

Die von den Astronomen des Mount Wilson Observatory gefundenen gesetzmässigen Verschiebungen der Spektrallinien der Nebel sprechen in der Tat für eine auch in Grossen nicht-statische Auffassung des Raum-Zeit-Kontinuums.

Mt. Wilson

A. Einstein Archive
23-467

PLATE 22. Typed draft of reply to inquiry from newspaper regarding the red shift observations carried out at Mount Wilson.

FIGURE 6.17. Joseph Weber with one of his gravitational-wave detectors. University of Maryland, ca. 1965. Special Collections, University of Maryland Libraries. Copyright 1969 University of Maryland.

the use and development of high-performance computing facilities necessary for simulations and for the evaluation of measurement results.[99] The discovery of a double pulsar losing energy through the emission of gravitational waves in the 1970s and its challenging theoretical analysis was a first culmination of the golden age that followed the renaissance of general relativity.[100]

In summary, some of the crucial steps in the investigation of gravitational waves were taken long before the recent detection of these waves by terrestrial measurements and long after Einstein had derived his field equations. A crucial turning point was the renaissance of general relativity beginning in the mid-1950s. This renaissance was not just the result of Cold War technology and funding as it may appear at first glance. It was brought forward by a community effort that explored the untapped potential of general relativity. In consequence, the theory was no longer conceived as a patchwork of isolated results or a stepping-stone toward a more final theory, but as a comprehensive physical framework in its own right. General relativity, from its creation by Einstein to its current advances, was a product of genius, but its transformation into a paradigm of modern physics was also, to a surprising extent, a community effort from the very beginning.

FIGURE 6.18. In September 2015, gravitational waves, generated by a collision of two black holes, were detected by the twin Laser Interferometer Gravitational-wave Observatory (LIGO) detectors, operated by Caltech and MIT, confirming a consequence of the general theory of relativity established by Albert Einstein 100 years earlier. Drawing by Laurent Taudin.

7

The Detection of Gravitational Waves

A Reflection

HARRY COLLINS

I started looking at gravitational-wave detection physics in 1972 and have been at it on and off for forty-five years. I am a sociologist, and during nearly all of that time one of my overarching concerns has been to feed into my sociological analyses an understanding of the science of gravitational-wave physics which compares with the way the physicists understand it. My aim has been to begin my account of what happens in the science from the perspective of a member of the community even if I depart quite markedly thereafter. Generating sociological analyses depends on estrangement from the insiders' perspective, but my view has been that understanding should come before estrangement.[1]

Here I try something different: an adventurous kind of history. It *starts* with estrangement—distanced reflection on what led up to the 2015 discovery and reflection on the meaning of that discovery; it is very much a view from outside, albeit based on technical understanding.

What I am going to be engaged in here, at least at the start of the argument, is what historians call the method of the "counterfactual": I am going to guess what would have happened if certain other things did or did not happen. There is no way of proving counterfactual claims, and the best one can do is put forward some plausible-looking arguments. This isn't the same as saying whatever takes your fancy, or presenting "alternative facts," because plausibility is a condition and it has to be established. That said, when it comes to counterfactuals, fools rush in where angels fear to tread. But if not now, when?

Fools Rush In

The idea of gravitational waves began around one hundred years ago with Albert Einstein. As Daniel Kennefick documents in his book, *Traveling at the Speed of Thought,* Einstein kept changing his mind about their existence, but at the moments that he did believe in them he also believed they were far too weak to be of any possible significance for physics. I think the idea that they were too weak to be of significance for physics changed with Joseph (Joe) Weber, who designed an apparatus for detecting them and went ahead and built it. Not only that, Weber then said he had detected signals that were consistent with his having detected gravitational waves. The high point of his claims occurred in the early 1970s. By about 1975 there was hardly anyone left who believed him. Weber died in September 2000. My view is that even though his claims were wrong—as nearly everyone now agrees—without Joe Weber we would not have detected gravitational waves to this day.

To emphasize the importance of Joe Weber, I am going to put forward two counterfactual claims embedded within one another. The first is that without Weber the Laser Interferometer Space Antenna—LISA, or something like it, would have given us, or be expected to give us, our first direct sighting of a gravitational wave. LISA is currently due to be launched in 2032. Without Weber, I am claiming, space-based observation would, today, be the only method being actively pursued for observing gravitational waves *directly*.

Gravitational waves were "indirectly" observed in the early 1980s through the measurement of the decay of the orbit of a binary pulsar by Taylor and Weisberg.[2] Without Weber, I am suggesting, over the last four decades all the attention with respect to detecting gravitational waves directly would have been on satellite detectors spurred on by what Taylor and Weisberg had found. This is because a space detector like LISA should be far more sensitive than an Earth-based detector—it *ought* to work, whereas Earth-based detectors ought not to work—and we would know what space-based detectors were looking for, whereas the sources for possible ground-based detection were speculative. A space-based detector like LISA looks for long duration, long wavelength, signals that can be integrated over an extended period and it is this that makes it so sensitive. Its location in space also isolates it from the noisy terrestrial environment and enables the interferometer to work over a long baseline consistent with the long wavelength of the long duration signals that come from every binary star system in the heavens. Such was the signal detected by Taylor and Weisberg, who fortuitously found a binary system one component of which was a pulsar that could be used as a clock; LISA-like systems don't need pulsars. I think it was Weber who drew attention to detection of short "burst" signals with terrestrial detectors and away from long duration signals and started astrophysicists thinking about the possible sources for such signals, such as supernovae and the momentary death spirals of binary systems. One can see the hopelessness of the short burst search: compare looking for the signal from a universe full of binary stars with looking for signals from the last second or less of their lives; there are obviously going to be incomparably more of the former and we already know they are there. But if you are stuck with building a detector on Earth, with a short-baseline instrument, these rare and dubious signals are the only kind you can see. They will have a short wavelength and a short duration and they might or might not be there and, if they do appear, they will be mixed up with the countless similar noises that come from our world.

A second counterfactual hypothesis embedded in this one is that the space-based detection of long duration gravitational waves would be much further advanced than it is today because attention would not have been directed to bursts of waves. Perhaps satellites would already be in place observing the long duration waves, or LISA would be scheduled for an earlier launch.

These suggestions are based on the logic of the thing and take no cognizance of, for example, the details such as who was making the decisions or the technical and financial considerations they brought to bear. NASA provided assistance at the beginning to the first of these space-project proposals known as LAGOS (Laser Antenna for Gravitational radiation Observation in Space) which was put in motion in the early 1980s, led by Peter Bender and Jim Faller, but the initiative faded in the face of budget cuts. It would be revived later with the interest of the European Space Agency.[3] I also understand that things went slowly in part because NASA and ESA were unconvinced by the technical possibility of the "drag-free" satellite design. This changed recently with the success of the LISA-pathfinder prototype.[4] But here I am ignoring this kind of claim, instead taking it that without Weber, people would have found different kinds of pathways more attractive, different kinds of arguments more convincing, and would have made different kinds of decisions in a very different world. If Joe Weber had not already populated the world with the concepts and activities associated with short duration, short wavelength gravitational waves, pushing forward in the search for long duration, long wavelength waves would have seemed more attractive, and people would have been more willing to force their way over what were seen as the technical hurdles; the main impetus to the search—the demonstration of what was to be looked for—would have been provided by Taylor and Weisberg, not by Weber.

To repeat, space-based detection of gravitational waves seems like a pretty reasonable project—it ought to work—whereas Earth-based detection is completely insane and in a reasonable world would not have been started for many years yet, certainly not before the possibility of direct detection of the long duration waves had been confirmed by an array of satellites. Luckily, at least for those contributing to this volume, we do not live in a reasonable world all of the time.

I have to admit that it is hard to maintain my second hypothesis, about the speeding up of the schedule for LISA or its equivalent, in the face of a recent email exchange with one of the founders of space-based detection, Peter Bender. He assures me that his own interest was triggered by Weber's claimed discoveries, and he thinks the whole space-based activity was given its impetus by Weber rather than Taylor and Weisberg. But even if I am wrong on this point and the implications of Bender's remarks are right, my central claim still survives: without Weber we would not have the 2015 detection. It may even be possible that there would be no gravitational-wave detection activity at all! Regardless of which, it suggests that science is a stranger thing than it is usually thought to be; here we have a widely acknowledged and widely scorned failure often taken to be a great embarrassment to physics, without which there would be no success!

Making Facts

If the arguments I am putting forward have any plausibility at all, it is because they are based in a sociological view of scientific change. That view is that, like every other kind of social change, scientific change happens because people start thinking *and doing* one kind of thing rather than another. This seems like a banal truism, but it takes us away from the idea that scientific change is monotonic progress that starts with ideas and works through theories to experiments or observations—from thought to action. What it says is that ideas are of no deep significance unless they are acted on; ideas do not have any meaning unless that meaning is evinced by what people do in consequence, and what people do affects what counts as a good idea.

This sociological theory of scientific change has a bearing on the relationship between theory and experiment but, in itself, it is not such a theory. It is not being claimed here that a theory has no reality until it is matched by experiment. Goblins, mermaids, and witches do have reality in the societies in which they are a determining feature of everyday life, and they have reality even in our society to the extent that they are supported by the activities of Hollywood and the publishers of fairy tales; something similar applies for theories in the absence of experiments. In that first fifty years of their life, before there were any corresponding experiments or observations, gravitational waves had reality to the extent that there were conferences, and conference proceedings were published, and all the activities associated with these things were taking place. But one can see that the reality of a scientific idea grows with more kinds of activity, such as observational and experimental activity, because more members and more elements of our society are drawn into working with the idea—more kinds of people's intentions and actions are informed by the idea and the idea becomes more real for us in consequence.[5]

One source of these ideas is Peter Winch's little book, *The Idea of a Social Science,* which was written in 1958. Winch, drawing on Wittgenstein's *Philosophical Investigations,* offers the germ theory of disease as an example of the relationship between concepts and actions. In my words, what he says is that when you see surgeons ritualistically scrubbing their hands and robing their bodies in sterilized garments before entering the operating theater they are, in effect, creating germs as much as removing them: through their actions they are making germs real for us. Their actions show us what germs are. A whole set of additional human activities and industries—firms that make sterilizing equipment and garments that can be sterilized, firms that make chemicals to destroy germs, firms that make antibiotics, firms that make microscopes and other laboratory equipment, firms that publish books and papers that describe germs, and endless works of fiction, including films and television, as well as works of history—continually create the meaning of germs. They create the meaning of germs just as the stories and ritualistic trials in certain places create witches or in other places create the transubstantiation of wine into blood and so on. Someone might have dreamed up the idea of germs, but without all these changed activities—say if surgeons were still operating in bloody waistcoats—then there would be no germs in our society, at least not to any greater extent than there are goblins or mermaids in our society. To repeat, an idea is not real unless it is embedded in the intentions and actions members of a society ordinarily perform, and that is why societies have such radically different cultures. As Wittgenstein put it, trying to solve the problem of how language gets its meaning, the meaning of a word is its use.[6]

In the standard stories of science, the direction is always from idea to activity but, to repeat, the sociological view I am putting forward here sees the two things as two sides of the same coin with causation running both ways. What I have been exploring in my studies over the years is how gravitational waves have gained a meaning in our society as a result of the uses they have found. What Joe Weber *did*—the way he used the idea and got others to use it—helped make gravitational waves real for us, and still more, what he *did* made the detection of gravitational waves *with terrestrial detectors* a reality within our society even if he did not actually detect any waves. Note that though gravitational waves may have meant something, although not a huge amount, in our society during the half-century of conferences and publications up to the 1950s, short duration gravitational waves of the kind that could be detected with terrestrial detectors meant nothing at all during that time.

One can spell this out in more detail. By building a detector or two and then claiming to have detected the waves, or signals consistent with the waves, Weber caused others to build detectors to check or refute his claims.[7] Because of this, funding agencies began to fund numbers of Earth-bound gravitational-wave detectors to check his claims—funding such detectors became close to, or closer to, an ordinary part of the funding agencies' day-to-day activities. And then others built more elaborate detectors—the cryogenic bars—which cost a lot more money and took a lot more effort to build and refine, so that this terrestrial detection activity became a lasting activity and a fairly expensive one. Consider: would anyone have built cryogenic bars without Weber? In theory, the cryogenic bars were nearly as unlikely to see any waves as Weber's bars, so why would anyone have built such expensive, complicated, hard to refine, and unlikely to succeed devices had they not had something to prove or disprove—something provided by Weber? And then, of course, the interferometers had something to prove or disprove in their turn; the feel of things was that the interferometers had to be built to do things properly.

As an idea, the interferometers began as an independently conceived way of detecting gravitational waves different from the resonant bars thought up by Weber. Weber himself came up with the interferometer idea and so did two Russian physicists, Gertsenshtein and Pustovoit, while the first such device for detecting gravitational waves was built by Bob Forward, one of Weber's team, but he was getting ideas from a certain Philip Chapman who, in turn, was in contact with Rai Weiss. Weiss first worked out what would be needed to build an interferometric device that could detect the kind of waves that theorists thought might be found, so though Weiss was not the first into the interferometer game, he was the first to think it through. But Weiss, as he readily admits, would not have been thinking this way were it not for Weber because it was Weber who caused theorists to conceive of astrophysical events that might give rise to the kind of short duration signals he said he was seeing. As Weiss, who never believed in Weber's findings, explained to me in 1975:

> Weber turned out to have been right in one regard, namely, there seems to have been a reason to look for fast events. I would never have thought of that. Never in my furthest mind would I have thought that you should look for things—that in astrophysics there could be events that would take milliseconds. And I would always have thought you'd look for something that was slow—because everything is slow . . . and so it fell right in with this business of the black holes, and the possible discovery now of black holes by X-ray astronomy.

So Weiss, who would otherwise have been thinking about detection of slow events of the kind that can be detected by the space-based LISA, was set on the road to terrestrial detection by Weber.

Magnificence Grows Out of Absurdity

A second, related, reflection, is that the really interesting changes in our way of being— the really striking social changes—are big. A big change is something that seems absurd when it is first proposed. The new view of the universe that Einstein came up with was "absurd" and was heavily resisted; indeed, it is still resisted in places. What could be more absurd than that a person can stay young just by traveling very fast, or that things get shorter the faster they go? But Einstein's ideas slowly gained ascendancy as more and more intentions and actions came to embody them. Einstein's ideas are "great," now that they have a social reality, because, when first put forward, they were absurd. Given the now sedimented "reality" of Einstein's ideas, LISA is reasonable. On the other hand, what Joe Weber did was absurd in spite of Einstein's ideas, not because what he said he had seen was incompatible with the theory of relativity, but because the numbers were ridiculous. Gravitational waves are so weak that the idea of detecting them on a noisy Earth is ridiculous and even the 2015 detection was ridiculous until it was announced. Extraordinarily, it was announced to almost no opposition, and almost immediately absorbed into everyone's commonsense via the mass media.[8]

Before 2016, a very large proportion of the scientific community thought the whole enterprise of terrestrial detection of gravitational waves was a waste of everyone's time and money. There was organized opposition to the first funding of the LIGO devices, opposition to the initial building of two detectors in the United States rather than one, and continuing skepticism and scorn throughout their development right up to the final moment. And in case of any misunderstanding, the expectation for Advanced LIGO was that, with luck, there might be a contested signal from an inspiraling binary neutron star around 2018; inspiraling binary black holes were not known to exist, though there was speculation that such a thing might be the first source to be seen. LIGO should never have been funded if the thought is that it is better to do the sensible thing than the absurd thing.

The magnificence of a scientific achievement, I'm suggesting, is, in large part, the extent of its initial absurdity—because that indicates the magnitude of the social and cognitive change it has to bring about. The magnificence of what was finally accomplished in 2015 was in the detection of the almost certainly undetectable: the equivalent of hair's breadth changes in the distance to the nearest star; the equivalent of a change in the diameter of the Earth of a proton's width; the actual measurement of changes in the length of a 4 km interferometer arm down to around one ten-thousandths of the diameter of the proton.[9] That these kind of distances would have to be measured was cited in evidence against the funding of LIGO. So if we are thinking about the relationship between theory and experiment, quite aside from the idea that meaning equals use, the two big social changes involved in the detection of gravitational waves were, first, Einstein's theory, and second, the actual accomplishment of the measurement. All the very clever theorizing that took place between the early years and the middle of the twentieth century and beyond, which was a necessary condition for the continuation of the experimental program, was not absurd. It was the working out of a predictable program which would not be expected to give rise to any counter-commonsensical understandings of the universe but would give rise to counter-commonsensical achievements in our ability to control it. To repeat, from the middle of the twentieth century onward, at least since the reality of black holes was accepted, the absurd thing—the thing that significantly changed our way of being in the world—was the ability to make these lunatic measurements, not to work out what might be seen if the measurement could be made.

The story told here, being driven by this thesis, mentions very few of the 1,200 or more scientists who were involved in the first detection. What I am looking for are certain decisive transition points, or crazy decisions, not the huge quantity of work done by enormously talented people that made the discovery possible; I am looking for things that turned the impossible into the conceivable. Weber, I have argued, is one such contributor because he changed reality; according to my thesis, without Weber the world of gravitational-wave detection would be a very different place.

This approach is sociological heresy in another way: sociologists generally deal in larger patterns, assuming individuals are replaceable—"if this person had not done the thing, social forces would have made it that someone else would have done it." The attribution of fame and prizes to individuals is sociological anathema. But my justification here is chaos theory; if the flap of a butterfly's wing can cause a tempest, then it is not unreasonable to think that what individuals do can create major changes as long as the conditions are just so. I am arguing that social forces would have led us to look for long duration gravitational waves rather than bursts had it not been that Weber's obsessions and actions fulfilled the role of the butterfly's wing.

More Than One Social Change

To go back to social forces and explain the position in yet another way, note that in terms of the science alone, much more was accomplished as 2015 turned into 2016 than the detection of a gravitational wave. First, it was discovered that there are black hole binaries in what are called the "intermediate mass" range—tens of our suns— and that pairs of these were born close enough together as the universe formed for there to have been time by 2015 for them to inspiral and merge. These things had been thought about, but neither of them was known before September 2015. Second, gravitational-wave astronomy has been initiated. Without the theory of sources that has been developed over the years from the middle of the twentieth century, we would not know about the intermediate mass black holes, nor would we have gravitational-wave astronomy. These things depend on the development of a "template bank" with which signals can be compared; the comparison indicates the origin of a detected waveform. Without the theory and the calculations, we would not know that in the 2015 event, one black hole was thirty-six solar masses and the other was twenty-nine solar masses, and we could not look forward to identifying the many other sources that will be detected. Identifying these will give rise to a new astrophysics and a new astronomy. But none of the theorizing that has gone into the construction of the template bank and the estimation of parameters was counter-commonsensical or absurd, it was just difficult—for instance it was not until 2005, after attempts lasting many years, that Frans Pretorius succeeded in simulating the inspiraling and merger of two black holes. These theoretical developments were part of the social change that has accompanied the detection, but not the initial cause of that change, in our society. It seems to me that the two kinds of accomplishment—the first detection and the foundation of a new branch of astronomy—are so different that they should be thought about in different ways. For example, they might well deserve two lots of, say, Nobel Prizes rather than one.

How It Was Done

Going back to the absurd detection, the third thesis is that to make ridiculous measurements requires something close to lunacy in the first instance followed by sanity to bring the lunatic inspirations to fruition. By lunacy I mean the kind of obsession that leads a person away from the normal run of human existence. Such a person chooses to do near impossible things that a normal person, even a normal scientist, would not choose to do. Without the lunacy no one would have tried to detect bursts of gravitational waves with terrestrial detectors, nor would they have invented the means to do it, while without the sanity they would have gotten no further than inventing—the inventions would not have succeeded. In more regular vocabulary, the two phases were realized in the form of small, unconstrained science followed by big, organized science. Of course, this being experimental science, and therefore needy of resources, the lunacy would have gone nowhere in the absence of certain entrepreneurial decisions; once it was given an initial impetus by Joe Weber, people had to be ready to take things further, and these people had to be not only scientists but enthusiasts with some influence about how resources get used. One generally overlooked contributor was Rich Isaacson, who supported and guided the program of terrestrial gravitational-wave detection through very difficult times when a more sensible occupier of his role at the National Science Foundation would have abandoned it, attracting a lot of applause. Kip Thorne, of course, had the enthusiasm to want to bring a piece of the program to Caltech, and Caltech soon took the lead in the American program as a result of his efforts and supported LIGO through difficult times.[10]

Per my argument, Joe Weber changed reality in such a way that entrepreneurs and inventors could find roles that had a chance of success in the enterprise of Earth-based gravitational-wave detection. The main inventors were Rai Weiss and Ron Drever. Ron Drever died in March 2017 after a period of dementia. I quote footnote 440 from my 2004 book, *Gravity's Shadow*, which no one seems to have contested so far; at the time I wrote it I didn't think it was of any great importance, but the discovery has made it of more moment and may well cause people to want to point to such mistakes as I acknowledge might be there:

> This is a retrospectively constructed list—it is a scientist's accounting rather than a sociologist's analysis. Thus, I count the Fabry-Perot idea as a success for Drever; Weiss wanted to use a "delay line." Drever won the argument, and most very large interferometers now use his idea. Perhaps the delay line will make a comeback one day, but for the time being the proper accounting of contributions for the purposes of this chapter is best done without moving far from actors' categories. Risking the wrath of angels, it seems to me that Weiss's main contributions were an inspired and inspiring working out of the consequences and method of implementation of the principles of experimental design outlined by his mentor, Robert Dicke, whereas Drever's inventions emerged out of what is sometimes called "lateral thinking."
>
> For scientist-readers, my rough list of inventions includes concentration near the dark fringe and the locking of the signal to that fringe, with the control signal needed for the locking as the output signal. These two interrelated ideas as well as the whole conception of how to achieve the kind of sensitivity needed in such a device belong to Weiss. Drever invented the use of the Fabry-Perot cavity and invented and codeveloped the Pound-Drever-Hall method for stabilizing it; he invented the idea of multiple interferometer beams in the same tube and the half-length interferometer realized with a midstation; he deserves at least half the credit for inventing power recycling and turning it into a practical proposition (Thorne helped

him with the calculations), and he deserves considerable credit for inventing an early notion for "narrow-banding" an interferometer—"resonant recycling"—which became the very important and practical "signal recycling" in the hands of others, notably Brian Meers, who worked in Drever's Glasgow laboratory. Drever may also have first put forward the idea of what is now known as "wavefront sensing." The list of inventions might well be longer, but about these there seems little doubt.

These are the big discrete ideas. Scots, Germans, Italians, and Australians have invented radically new kinds of mirror suspension, and a still more advanced form of recycling called resonant side band extraction was invented by a Japanese graduate student, Jun Mizuno, while working in Germany. Innovations have also come from US LIGO Scientific Collaboration groups, notably a way of attaching mirror supports, which was invented at Stanford University. The Russians have led the way in new materials and new theoretical understandings of the devices. We also should not underestimate the value of more recent contributions to the integration and execution of LIGO, without which these ideas would have been stillborn. As one respondent put it, it is vital "that the engineering challenges are met, and that the design and implementation evolve during the design and building to optimize the product, as well as meet cost-schedule constraints." What I have said in this footnote almost certainly contains mistakes. (p. 556)

For those who do not know the story, in 1984, the LIGO project was created by merging two teams of interferometer scientists, one led by Weiss at MIT and one created by Kip Thorne at Caltech but headed up by Ron Drever, who Thorne recruited from Glasgow University.[11] Drever had been building gravitational-wave detector resonant bars and then interferometers. Drever developed the big, 40m prototype interferometer at Caltech. In July 1992, however, he would be sacked from the LIGO project and no longer allowed access to the prototype. The "structural" reason for his being sacked, according to my analysis, was that though he was somewhat of a genius at inventing, he was no good at working in a team, and he certainly could not work in the kind of team needed in the transition to big science with its organization and continual compromise; big science must often give priority to deadlines, sacrificing the possibility of finding the very best scientific solutions, and Drever found this hard to impossible. The details of Drever's departure were bloody, but they need not concern us here; they are largely set out in *Gravity's Shadow*, along with many more details associated with the story of gravitational-wave detection from the beginning until the early 2000s.

What is important is that before he left the project, Drever contributed vital ideas and design decisions. They are in the list above. Someone had to invent "power-recycling" if LIGO was to work, and it seems that Drever at least half-invented it and certainly pressed it upon LIGO. Signal recycling was also important, but Drever was probably not the main inventor of that technique. Perhaps most important was the Fabry-Perot cavity idea as the method of reflecting light backward and forward in the interferometer arms in order to lengthen the light path given the fixed arm length. He was not the first to use this idea in a big interferometer, but he was the person who saw its importance for LIGO and, earlier, he had co-invented the necessary method for controlling the mirrors in such a sensitive device—it is known as the Pound-Drever-Hall method.[12] In the matter of Fabry-Perot cavities, Drever was embattled with Weiss who wanted to use what are known as "delay lines," which bounced narrow individual beams from mirror to mirror in a zigzag path; the Fabry-Perot cavity reflects a full-width beam between two mirrors like the infinite reflections you see if you stand in a mirror-sided elevator.

Drever and Weiss were locked into irresolvable arguments about this and similar design features at the time LIGO was getting under way. This was when it was run by what became known as the "Troika"—a committee consisting of Weiss and Drever, chaired, ineffectually by his own admission, by Thorne. At this point LIGO was going nowhere, but after another, short-lived failure of a management reorganization it was taken on, in 1987, by Rochus (Robbie) Vogt.

Vogt was a strong leader and was the first to attempt to organize LIGO properly and make a realistic bid for funds. By 1990, funding had been agreed upon and by 1991, funds began to flow. Vogt's style, however, was charismatic leadership of a loyal team and, being a long-standing Caltech person, his team tended to consist of Caltech personnel with the MIT group led by Weiss marginalized. (See figures 7.1 and 7.2.)

As it happens, this was lucky because Vogt settled the big questions about the science in Drever's favor—Drever being a Caltech man insofar as he was anyone's man. That meant that Drever's Fabry-Perot method was chosen ahead of Weiss's delay line, and so were a number of other things favored by Drever. Someone who had no reason at all to favor Drever told me: "the interferometers we built were very much the ones Ron had proposed in 1989, not what Rai had proposed."

FIGURE 7.1. The Troika personnel in early days—Weiss, Thorne, and Drever. (Left) Reproduced with permission of Rainer Weiss, (middle) courtesy Caltech, (right) American Physical Society.

FIGURE 7.2. Thorne, Drever, and Vogt with the 40m interferometer. Courtesy Caltech.

At least some of these choices were what made the 2015 detection possible.

Like any counterfactual, even when it is a matter of scientific technique, it is possible to argue about it and I have heard it said that the delay line could have been made to work. But we know that the Fabry-Perot cavity design allowed for smaller mirrors and solved some other problems. On this detail, my method of reaching a plausible conclusion was to ask one of the scientists who (a) does not think much of

Ron Drever, (b) really likes Rai Weiss, and (c) knows as much about the technicalities of designing and building the machine that detected gravitational waves as anybody, and he said:

> I think that it would have been very difficult to get to where we are, and to think of a path to the future, using Delay Lines. It has to do with the size of the mirrors and the thermal noise issues.

So, although I cannot *prove* the claim to the standards of a physical science, it is highly probable that Drever's ideas, and the short period in the development of LIGO when Vogt was choosing Drever's ideas over Weiss's, was necessary for the 2015 detection. The delay line had a long history and had been used in a prototype gravitational-wave detector in Germany (figure 7.3), so using Fabry-Perot cavities instead was a kind of lunacy, but it was the right kind of lunacy.

As intimated, the period when Drever's ideas were being actualized in LIGO's design did not last long, but it seems to have been a crucial period. There would have been no design at all without Weiss, but there is reason to think that if the design had been left entirely to Weiss it would not have worked. So, insofar as we are allowing ourselves to assign society-changing effects to individuals, we can say that in scientific terms, the crucial lunatic contributions up to now in the history of gravitational-wave detection were made by Weber, Weiss, and Drever, with Vogt also playing a crucial part in the scientific decision-making; Vogt also began to bring some sanity into the detection effort. Of course, without Isaacson there would not have been much to discuss and without

FIGURE 7.3. Light paths in a delay line. This Munich-based device using green light with 47 bounces off the end mirror. Courtesy of Albrecht Rudiger.

Thorne's entrepreneurial enterprise there would have been no Caltech interferometry project and so no LIGO.

In 1994 Vogt was replaced as director. As indicated, in structural terms, Vogt had made the first moves to shift terrestrial gravitational-wave detection from lunacy to sanity. For example, in addition to supporting Drever on many technical details, he also cut right back on Drever's plans to have a half-dozen interferometers in each beam tube, instead accepting Drever's fallback idea of a second half-length interferometer installed at Hanford, which ought to confirm any sighting on the main interferometers with a half-size signal; as a result Vogt turned LIGO into a practical proposition. But Vogt would not take more than the initial step.

First, Vogt wanted to run LIGO like a "skunk works"—people throw money over the wall and you and your loyal team deliver the goods after a period of years. So Vogt would not provide detailed plans and schedules even though the NSF needed them both to assure them that money was not being wasted and to assure the wider and wider group of parties, who find they have an interest when a project starts spending hundreds of millions, that nothing irresponsible was going on. LIGO was a very unpopular project, with many scientists who might apply to the NSF for research funding for small sums seeing LIGO as limiting their chances of success; this constituency could not be alienated further by a project that appeared to refuse to account for itself. Second, Vogt's charismatic style was not suited to expanding the team in the way needed to spend big money and draw in scarce talent scattered across the nation and the globe; Vogt was already alienating members of the MIT team which was *part of LIGO*! His approach was not designed to recruit large numbers of scientists, whose first loyalty was elsewhere, to be enthusiasts for LIGO, but that was what was needed.

In 1994, Vogt was replaced by Barry Barish and his chosen project manager, Gary Sanders. Barish and Sanders set about transforming LIGO further—it had a long way to go to become a proper, accountable, big science project, but Barish and Sanders set about taking it there. Work practices were changed, with round-the-clock shifts and much more routine being introduced. The way the science of interferometry was thought about was also transformed; it was now conceived of as something that was sufficiently well understood to merit building full-size devices. Up to now interferometry had been considered mysterious, resting on continued inspirational work with prototypes to understand the noise. Put simply, Barish and Sanders decided that the recalcitrant problems of noise in the prototype were due to lack of standardization in the prototype rather than anything deep, and that the noise was now sufficiently well understood to merit trying to solve the remaining problems on the full-scale devices rather than on unrepresentative models. This was a gamble, and the experts from Vogt's team assured Barish that there were still unknown sources of noise that made such a transition premature; there were many doubters, even among the insiders, when the devices were first switched on. I well remember those moments and, reflecting opinion among the community, I remember that I was quite unsure about what was going to happen. But in the long term, Barish turned out to have made the right decision.

This change of direction, with the shift work imposed by Sanders, led to resignations by many of the key players from Vogt's team and the resignation of Vogt himself. Under the Barish-Sanders style of management, however, the departing interferometer experts could be replaced by others from the worldwide community and the resignations did not cause the major problems that might have been expected. Crucially, LIGO now became accountable and this meant it could survive.

The complaint by those who left was that Barish was turning the project from something designed to detect gravitational waves—which still required creativity so as eliminate sources of noise that were not yet understood and still needed much more sensitivity—to something designed to reach a predicted level of sensitivity that would not be enough to detect the waves. But inasmuch as this was true, which it probably was, even though it could be argued that Vogt's project was already aiming for only this level of sensitivity, it was the right transformation to make. Only this way could clear targets and timetables be set and accomplished (or near-accomplished) and only this way could accountability be delivered to those in the long opportunity-cost shadow of this most expensive ever NSF project. Under the Barish regime LIGO was made to work under a restricted definition turning on promised sensitivity, and this level of working was

FIGURE 7.4. Barry Barish and Gary Sanders when they took over LIGO. Reproduced with permission of Barry Barish.

sufficient to justify further funding. It would eventually justify the funding of Advanced LIGO even though Initial LIGO had not detected any waves. The justification was the manifest meeting of promises accomplished via managerial virtuosity. It would, of course, be Advanced LIGO that would make the magnificent discovery.

Summary and Conclusion

This may seem an odd place to stop reflecting on the scientific story since from the end of the Barish/Sanders regime in 2005 to the discovery of gravitational waves, there followed ten years of intense and successful work. But this was not work that required either transformation of the reality in which we lived or any kind of lunacy. The transformation in our social lives may not have been secured until 2015, but the way forward had been established by Initial LIGO. That is why I merely mention that Jay Marx became the director in 2006 and that when he resigned in 2011, Dave Reitze took over and that David Shoemaker very successfully oversaw the development and building of the Advanced LIGO that would make the first detection, on schedule and within budget.

I have argued that the detection of gravitational waves in 2015 represents a step in a major social change which has made the ability to make what were once impossibly small measurements into something routine; it is now rapidly becoming one of the day-to-day activities belonging to the short-duration-gravitational-wave astronomy of the future. The measurements are so small as to be absurd and were certainly thought absurd by a large proportion of the science community who were in a position to understand what was proposed. This was the social change that caused the 2015 detection to be such a magnificent scientific achievement. All those who thought it impossible have had to accept that they were wrong, and it is pleasing that many of them are finding ways of saying so. Scientific society is changing rapidly as the absurd becomes ordinary and new ways of going on are building up fast. As we have seen, the ability to accomplish this change required a reality-changer—Joe Weber—who was prepared to try to do the impossible and say he had done it; it required inspired inventors, who thought of ways in which what Weber said he had done could be done; it required transformers who turned the ideas of what I have called the "lunatics" into a doable, and crucially, *fundable*, science; and it required entrepreneurs and people ready to take a risk on the funding.

8

Einstein at Caltech

DIANA KORMOS BUCHWALD

When Robert A. Millikan became the director of the Norman Bridge Physics Laboratory at Caltech in 1921, where he would serve as de facto president for more than two decades, he embarked on a systematic program of bringing renowned European scientists, such as Einstein, to this budding scientific and technical research endeavor, the first of its kind in Southern California.[1]

Einstein's work on light quanta dating back to his 1905 paper on the photoelectric effect[2] found subsequent support when Millikan confirmed Einstein's relation between energy and emitted frequency.[3] Einstein's personal interactions with Millikan began during his first trip to the United States in the spring of 1921, when he was introduced not only to a large community of East Coast Jews among whom he was raising funds for the establishment of a Hebrew University in British Mandate Palestine, but also to physicists and astronomers. Before his arrival, Einstein had written to an acquaintance in Chicago that he wished "to come into contact in so far as possible under the circumstances with the American World of Science, those with whom I have more common ground scientifically."[4] Among them were the scientists working at the Yerkes Observatory, which housed the largest telescope in the world at the time, built by Hale in 1892 some 100 miles northwest of Chicago. By the time Einstein visited Chicago, Hale had already implemented the Yerkes model of an observatory that integrated observational devices with a laboratory atop Mount Wilson, above Pasadena, where the 100-inch telescope was soon to yield revolutionary astronomical discoveries.

Einstein's 1921 visit to the area was spectacularly successful, to the extent that he was offered the University of Chicago professorship soon to be vacated by Millikan. Einstein declined, even though that same year he had faced physical threats and anti-Semitic animosity in Germany: "It moved me very much" to be offered such a wonderful position, he wrote. "Although it is true that I had some ugly experiences with colleagues and students, I am so firmly rooted here by family ties and friendships . . . I could not choose a completely new, even if attractive, environment." In such a "completely changed milieu," he could not imagine himself capable of the necessary "regeneration" without incurring "significant damage."[5]

In May 1922, five months after the ratification of the US-Germany Peace Treaty, Millikan informed Einstein that he had been elected to the National Academy of Sciences, a gesture that Einstein appreciated as a sign of the gradual amelioration of international relations among scientists following the animosity of the war years.[6] Also in 1922, the revered Dutch theoretical physicist Hendrik A. Lorentz visited Caltech, where he lectured on the most important recent developments and outstanding problems in physics. Shortly thereafter, Millikan also offered Paul Ehrenfest, then Einstein's closest friend and colleague, a visiting appointment for one term, funded by the National Research Fellowship board. He would be asked to give six to eight lectures at Caltech as well as nineteen at other universities over a period of three to four months. Ehrenfest's remuneration would be $600 per month,[7] of which Millikan expected that no more than $150 were needed for monthly local expenses, plus steamship and rail tickets. Similar arrangements pertained to the visits of Lorentz and Henri Pirenne, Pirenne being "the foremost" European historian in Millikan's words.[8]

Official relations between US and German scientists were nevertheless still cool. Both Hale and Millikan had participated in the US scientists' war work, as had numerous Germans. Millikan traveled to Europe in spring 1922, and upon returning in the fall reported to Hale that he feared, "unless the rampant inflation in Germany stabilized, the world would soon see revolution there." He had had no conversations with "German scientific men except with Einstein" and the physicist Wilhelm Westphal, an expert adviser to the Prussian Ministry of Science and strong supporter of Einstein's relativity, "both of whom are of the thoroughly reasonable sort, and both working devotedly to bring about better understandings than now exist."[9] Following the successful Eddington eclipse expedition of 1919 and the public controversies surrounding him on the German scientific and political scene, Einstein was being offered appointments elsewhere. Paul Ehrenfest and H. A. Lorentz made him a part-time visiting professor in the Netherlands, while Columbia University, stimulated by Millikan's report concerning Einstein's "predicament" in Germany, offered a full-time position, which again Einstein declined.[10]

Einstein's relations with Caltech soon intensified. In June 1922, the Russian mathematical physicist Paul Epstein, whom Einstein had successfully recommended for a teaching position at Caltech in 1921, suggested to Einstein that the twelfth available seat on the newly established International Committee for Intellectual Cooperation of the League of Nations be reserved for Millikan. Since Hale's health was deteriorating, Epstein could think of no more illustrious an American scientist to hold that position. But Hale had already accepted a few days earlier, and Millikan was designated as his alternate on the commission.

In October 1923, while on a trip to Japan, and following, metaphorically and literally in Einstein's steps, Millikan was awarded the Nobel Prize in physics. He thanked Einstein for his letter of congratulations: "I suspect that I owe this award in no small degree to some kind words which you may have been guilty of saying in my behalf to the committee. In any case, I wish you to know that there are no congratulations which I appreciate more than those which come from the man who has contributed in such an extraordinary way to practically all of the developments of modern physics."[11] Einstein, who had been on his way to Japan a year earlier when he himself was notified of winning the Nobel Prize, had not in fact nominated Millikan. The prize awarded to Millikan was yet another justification for the earlier award to Einstein. That year, Einstein was also in touch with other Pasadena astronomers, including Gustaf Strömberg who, upon the request of Walter S. Adams, director of the Mount Wilson Observatory, solicited Einstein's opinion concerning papers soon to be published in *Mount Wilson Contributions*.

Upon returning to Europe following his visit to Pasadena during the winter term of 1924, Ehrenfest wrote Einstein:

> Millikan would very much like to invite you to Pasadena, same as Lorentz and me, and asked me to probe whether he could get you.—Well, I'd like to tell you the following about this: the 3 months in Pasadena were completely splendid for *me*—during the winter the climate is quite wonderfully pleasant, cool air throughout, very warm sunshine.—The people *very* nice—nothing ugly all around (whole city full of trees—very beautiful untouched nature in close proximity, conveniently reachable by car).—*I'm yearning* to go back there very soon. I love California *very much*, whereas eastern America, although quite interesting, not *at all* attractive to me.—There was absolutely no rushing about for me in California, but there certainly was in the East. It would be particularly fun for me to be able to be in California

> once with you!! I don't quite know which year Millikan would like to have you—in any case, definitely choose January, February, March for Pasadena, not later in the year!![12]

Einstein responded with reserve: "I congratulate you on your happy return and honestly admire that, after touring around for so long now, you can already think of going to California with me. For me it's the other way around, however. As I'm a bit unsociable, the thought of such a journey is quite unsettling, beautiful though it may be over there. Moreover, I promised the South Americans that my next trip outside Europe would be to go to them. I should have gone there this year already but couldn't resolve to do so. . . . So tell the people (that is, Millikan) that I find it very kind of them to have thought of me, but that I request they not invite me for the time being."[13]

Nevertheless, by October, after having again met Einstein in Geneva, where they probably discussed the matter, Millikan sent a 1,000-word official sounding invitation that says much about his ambitions for Caltech only three years into his tenure: "I am writing to ask if you would not like some time to follow the good example set by Dr. Lorentz and Dr. Ehrenfest and spend the months of January, February, and March playing in and about the Norman Bridge Laboratory and the Mount Wilson Observatory." Einstein was not expected "to give any courses at all, but merely to be one of the half a hundred men, all intensely interested in the latest developments in physics and astronomy"; he would be attending "our seminar three times a week, and could also get into contact with the very large amount of magnetic results which have been accumulated." Millikan wished to work with Einstein on "an important analogy between X-ray spectra and optical spectra," and promised that he would be able to work "undisturbed." He reassured Einstein that "We would keep the reporters away from you, and let your stay here be solely for the purposes of the progress of physics rather than for publicity purposes. We should simply ask you to be a 'Research Associate.'" The Carnegie Institution and Caltech would provide a stipend of $3,500, which "would pay all expenses of yourself and Mrs. Einstein from Germany here and back and leave at least a half of the stipend to be used for other purposes . . . Lorentz agreed to return again." He requested an instant reply: "a cable addressed simply 'Millikan, Pasadena' would do, saying 'can come' or 'cannot come' as the case may be. You will of course send this cable collect."[14]

Einstein telegraphed immediately: "Millikan, Pasadena, Cannot 1925, Einstein" and followed with a letter.

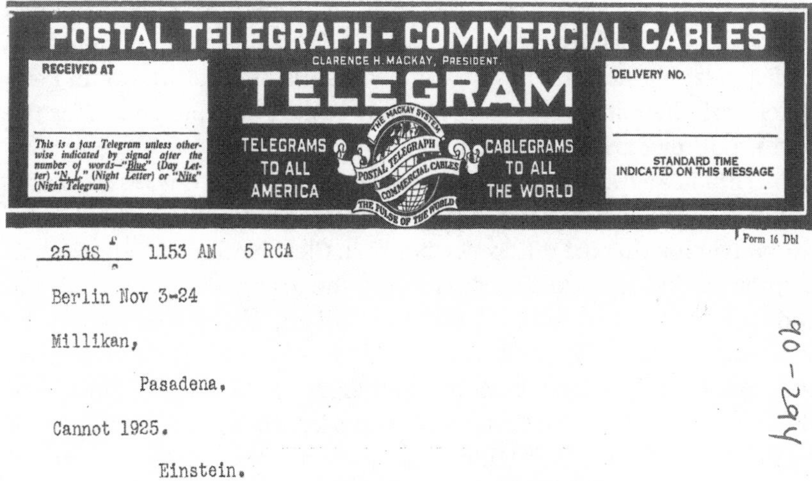
POSTAL TELEGRAPH - COMMERCIAL CABLES

CLARENCE H. MACKAY, PRESIDENT.

TELEGRAM

RECEIVED AT

This is a fast Telegram unless otherwise indicated by signal after the number of words—"Blue" (Day Letter) "N.L." (Night Letter) or "Nite" (Night Telegram)

TELEGRAMS TO ALL AMERICA

CABLEGRAMS TO ALL THE WORLD

DELIVERY NO.

STANDARD TIME INDICATED ON THIS MESSAGE

Form 16 Dbl

25 GS 1153 AM 5 RCA

Berlin Nov 3-24

Millikan,

Pasadena.

Cannot 1925.

Einstein.

90-294

FIGURE 8.1. Telegram from Einstein to Millikan, November 23, 1924.

Since Einstein planned to travel in 1925 on a promised lecture tour to South America, he could only come to Pasadena the following year at the earliest: "I thank you very much for your friendly letter and hope that I can come and visit you in 1926 together with Lorentz. If it should not be possible for me to travel in 1926, I would definitely come 1927. I already am very much looking forward to it. I would also very much like to speak with the gentlemen in Washington about the problems of terrestrial magnetism. But my thoughts in this direction are still quite indefinite at the moment." To Epstein, who reiterated the invitation, he stated he "would definitely come in 1927 at the latest."[15] But eventually he would not travel for five years after his return from South America in 1925.

A few months later, Einstein again received news of scientific work done in the United States, in particular new work by Dayton C. Miller of Case Western University, who was performing thousands of measurements that purported to prove the existence of an ether. Einstein had encountered Miller's previous experiments during his first visit to America in 1921 and had expressed serious reservations about their soundness. This time he was asked to comment on the proofs of Miller's latest papers. He replied that if Miller's results should be confirmed, "then the special relativity theory, and with the general theory in its present form, falls. Experiment is the supreme judge."[16]

Most physicists were equally skeptical. Epstein wrote to Einstein that "our circle accepts Miller's somewhat daring statements under great reserve; we hope we will be in the position to check his measurements with other kinds of instrument."[17] When asked, Einstein commented to colleagues that he disbelieved Miller's results but that he would keep quiet until the matter was resolved in the court of experiment. "At the bottom of my dark soul I don't place much weight on Miller's experiment, but I can't say it aloud," he wrote to Ehrenfest. "It is less a matter of blind confidence" in the theory of relativity "but rather because of the conviction that the difference between Cleveland and Mount Wilson cannot be that significant, considering the grand scale on which the Old One has created the world."[18] To Millikan, Einstein wrote that he was very curious to find out further details about the "counterattack" that had been launched both on the heights of Mount Wilson and at the lower altitude of the Caltech laboratories. He again confessed that he did not make much of Miller's data, which must rest "on sources of error. Otherwise the entire theory of relativity collapses like a house of cards."[19]

In January 1926, even though new data had not yet become available, Einstein was pressed by a European correspondent of the Hearst Universal Service to state publicly that he was ready to bet that Miller's experiments were in error:

> If the results of Prof. Miller's experiments should be confirmed the theory of relativity could no longer be maintained. For the experiments would then prove that, with respect to the movement of the earth under certain conditions, the speed of light in a vacuum would be dependent on direction.
>
> Therewith would be disproved the principle of the constancy of the speed of light, which is one of the foundation stones, and the one on which the theory of relativity rests.
>
> There exists, however, in my view, practically no possibility that Prof. Miller is right. His results are irregular and point more to a still undiscovered source of error than to systematic effect.[20]

By the fall of 1926, once the work of R. J. Kennedy at Caltech was concluded with "considerably higher sensitivity than D.C. Miller's,"[21] both Epstein and Millikan could confirm that no ether drag had been shown either in the laboratory or at Mount Wilson. Millikan took the opportunity to reiterate an invitation to Caltech: "I am hoping that your health is now so far improved that you may wish again to consider the possibility of spending a winter in Pasadena."[22]

Three years before, Einstein had revived work on the affine formulation of general relativity that had been advanced by Herman Weyl and Eddington. Einstein thought that the reformulation could yield empirical consequences that he was hoping to be able to test experimentally. As a result, they also bore implications for "an understanding of terrestrial magnetism and what he called the electromagnetic bookkeeping of the Earth."[23] This was a topic that had fascinated him since childhood, and in the early 1920s he devised experiments, hoping to simulate the currents of hot mass in the interior of the Earth. Though we do not know exactly what those envisaged experiments might have been, they involved ionized gases and space charge. Rumors that Einstein had a new theory of terrestrial magnetism appeared in the *New York Times* in 1923. Hale wrote in late 1926 that he was greatly interested in Einstein's new theory of gravitation and magnetism. He also informed Einstein that Michelson himself would "soon repeat the Michelson-Morley experiment," given the preceding two-year-long Miller affair.[24] Einstein was pleased by Hale's supportive letter, but also despondent: "If only it were true that I had a theory of the terrestrial magnetic field! Although a couple of years ago I did brood over a pertinent egg, nothing living hatched from it." And he continued: "I have already received so much kindness and great scientific support from Mount Wilson that it is increasingly difficult for me to always say 'no' to the best men there, who are repeatedly inviting me. But I would not be equal to the ordeals of such an expedition anymore."[25]

Despondent he may have been, but Einstein nevertheless worked with Lorentz, who arrived in Pasadena in January 1927, staying at the Strathaven Hotel, to make sure that Millikan would be invited to the upcoming Solvay Congress.

An important letter, preceded by a telegram, arrived in 1927 from one of America's most respected mathematicians. Oswald Veblen, who had heretofore corresponded sporadically with Einstein, now invited him to become a research professor at Princeton University. He wrote: "I have just sent you a telegram which you will surely regard as very extraordinary—and so I must write a few words of explanation." He went on to discuss Princeton University's two-year campaign to raise money "in support of research in pure science." As a result, a program to appoint research professors "free from the duty of teaching undergraduates, and whose sole duty is to advance their respective sciences" had met with some success. Veblen himself was appointed to the "first of these chairs (mathematics)," together with K. T. Compton (physics),

FIGURE 8.2. The Strathaven Hotel in Pasadena, where H. A. Lorentz resided in 1927.

H. S. Taylor (chemistry), H. N. Russell (astronomy), and E. B. Conklin (biology), all of whom were Princeton professors at the time. Veblen then left for a one-year trip to Europe, whence he was now reporting that Compton had sent a telegram "asking me to find out whether you would consider a research professorship in Princeton." Compton asked Veblen to travel to Berlin to discuss the idea, but Veblen was "already booked to sail from Southampton" on September 21. He explained that the position would be remunerated with a salary of $10,000 per year, and that Einstein's duties would be "to continue your researches and to give such help to advanced students as would be compatible with your own studies." Veblen himself was giving one lecture a week. He further explained that research conditions had improved considerably since Einstein had visited six years earlier. Nevertheless, he recognized that he was writing with "only a faint hope," since America seems far away for a European, and reassured Einstein that they would "do everything in our power to make it pleasant" for Einstein. He encouraged Einstein to come "for a year and see how you like life on the other side of the Atlantic."[26]

Einstein, who was ill at the time, replied—to the (non-extant) telegram, and not to the letter—on September 17, 1927: "Very early this morning I received your friendly telegram. It pleases and honors me that you wish to appoint me at Princeton. I thank everyone very much for your efforts. But nothing will come of this. When one has grown old somewhere, on should stay there forever; an old flowering plant should no longer be transplanted, as it would then wither prematurely."[27] He was only forty-eight years old. This invitation would constitute the seed of future developments.[28]

Two years later, in the summer of 1929, Millikan again offered Einstein a visiting appointment for winter term 1930; he proposed a stipend of $3,500, as had been given to Lorentz and Arnold Sommerfeld. But through "the generosity of a mutual friend," Millikan would now also cover all travel and living expenses for Einstein's wife and for a personal physician as well, if needed.[29] Despite the "difficult temptation" and "embarrassing offer," Einstein did not accept. Moreover, he added:

> I must also say that my formal studies of the last years, as interesting as they may be, have brought with it the fact that I have followed only incompletely the stormy developments of physical theory. On the other hand my own studies have not blossomed sufficiently to warrant my being certain of their physical fruitfulness. It is only a bet on the future. So you won't miss much by my not coming.

Elsa Einstein also wrote to Millikan that same day. Her husband had pondered the generous offer for many days, and had declined with a "heavy heart." She explained that Einstein was declining many other invitations, to Russia and England, for example. And he feared the many additional invitations and unavoidable obligations that would crop up in New York or Chicago were he to be known to travel to Pasadena. Einstein's health had improved over the past year, Elsa Einstein wrote, but he needed to be watchful. She fondly remembered being Millikan's wife's guest at a Chicago club in 1921, and how dexterously Greta Millikan had driven her around the windy city.[30]

Einstein's fame had risen with the 1919 solar eclipse expedition that confirmed light bending in a gravitational field. This was one of three major predictions of general relativity. But the new theory also required that a clock placed in a gravitational field will tick more slowly for an external observer than an identical clock placed in a gravity-free environment. Because each spectral line represents a clock in a radiating atom at the surface of a star, the slower the clock ticks, the lower the frequency of the radiated line will be. This effect appears as a shift of the spectral lines toward the red relative to its wavelength measured on Earth. For the Sun, the Einstein effect is exceedingly small, a wavelength shift of only 0.010 Å for a line at 5000 Å ($\Delta\lambda/\lambda \approx 2\times10^{-6}$). His best-known

wächst also mit dem Gravitationspotential. Hieraus schließt man, daß die Spektrallinien, welche auf der Sonne erzeugtem Lichte zugehören, gegenüber den entsprechenden, auf der Erde erzeugten Spektrallinien eine Verschiebung nach Rot hin aufweisen im Betrage

$$\frac{\Delta\lambda}{\lambda} = 2 \cdot 10^{-6}. \quad [51]$$

FIGURE 8.3. Excerpt from "On the Relativity Principle and the Conclusions Drawn from It," 1907, as reproduced in *CPAE* Vol. 2, Doc. 47.

numerical prediction of this effect dates from 1916, but it had already been present in his first paper on gravitation of 1911, where it was directly derived from the equivalence principle, and was envisaged even earlier.[31]

Carrying out the sensitive measurements required for the confirmation of the effect was difficult. Einstein followed with great interest the progress in observational attempts to verify relativistic gravitational redshift. He was happy with favorable results obtained in Germany as early as 1920 by Leonhard Grebe and Albert Bachem,[32] and hoped that astronomers elsewhere would also succeed. He advocated for and succeeded to have built the Einstein Tower telescope in Potsdam for just this purpose. As Allan Sandage wrote in his magisterial history of the Mount Wilson Observatories, "no program of detecting the relativity effect at the 0.008 Å level could be launched before absolute laboratory wavelengths had been determined."[33]

On September 8, 1921, the *New York Times* reported that astronomers were now expecting the theory to be confirmed by the latest work of Charles St. John, working at Mount Wilson, who had previously attributed observed redshift to other solar phenomena. Einstein inquired of Eddington about the validity of these rumors that had reached him from across the Atlantic.

So important were these new measurements that Ehrenfest tentatively planned to bring Einstein and St. John together in Leyden as early as 1922. As St. John was carrying out these extraordinary observations and data analyses, Edwin Hubble's work between 1923 and 1927, with his discovery of stars in nebulae, transformed Mount Wilson from being a site primarily for solar work into a major locus for observational cosmology. Together with Vesto Slipher and Gustaf Stromberg, Hubble published measurements for the velocities of forty-five galaxies that were so much higher than the values encountered for stars that "they clearly concealed a deep mystery." This was especially significant because Einstein's 1916 paper "with its revolutionary notion that space curvature had everything to do with gravity, promoted galaxies to the rank of celestial markers of the space, perhaps revealing the large scale geometry of the universe itself."[34]

In his 1917 paper on cosmology, Einstein had described a homogeneous, static, spatially closed universe. In order to counteract gravitational contraction he had introduced the cosmological constant.[35] Five years later, Alexander Friedman demonstrated that general relativity allowed for dynamical solutions as well, and in 1927 Georges Lemaître envisioned the possibility of an expanding universe. Before the late 1920s, Einstein had been skeptical of the possibility, as he had also been of Karl Schwarzschild's and Friedman's work on singularities—now known to exist as black holes. However, we do know that Einstein had discussed the matter with Lemaître already in 1927, and that he possessed a reprint of Lemaître's paper,[36] whose proposal was supported by observational work at Mount Wilson, work that was emerging publicly only very slowly, since Hubble was famously reluctant to publish quickly. Historians have debated whether Einstein was converted to the new view during his stay in California. However, interviews upon his arrival in the United States indicate otherwise.

On January 2, 1931, Einstein answered in German a number of questions posed to him by the Associated Press (see figure 8.4). His reply, modified by the journalists, published in the *New York Times* a day later, was unequivocal: "New observations by Hubble and Humason concerning the red shift of light in distant nebulae make the resumptions near that the general structure of the universe is not static. Theoretical investigations made by La Maitre [*sic*] and Tolman show a view that fits well into the general theory of relativity."[37] Einstein had not stated "new" observations.

And a few weeks later, upon returning from his first visit to Pasadena, he published a paper entitled "On the cosmological problem" wherein he explicitly referred to Hubble's work on an expanding universe. In May 1931, before the paper reached print, Einstein traveled to the UK, where the second of the three Rhodes Lectures he gave at Oxford University on May 16 concerned the cosmological problem.[38]

There is no extant correspondence in Einstein's papers with Caltech concerning practical arrangements prior to his arrival in Pasadena, but material in the Caltech Archives shows that in October 1930, Arthur Fleming, chairman of Caltech's board of trustees, began searching for

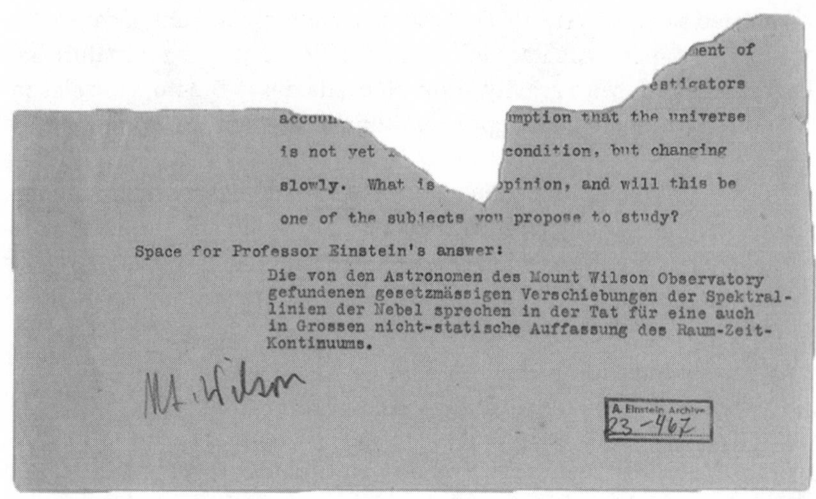

ment of

estigators

accoun mption that the universe

is not yet condition, but changing

slowly. What is pinion, and will this be

one of the subjects you propose to study?

Space for Professor Einstein's answer:

Die von den Astronomen des Mount Wilson Observatory gefundenen gesetzmässigen Verschiebungen der Spektrallinien der Nebel sprechen in der Tat für eine auch in Grossen nicht-statische Auffassung des Raum-Zeit-Kontinuums.

Mt. Wilson

A. Einstein Archive
23-467

FIGURE 8.4. Typed draft of reply to inquiry from newspaper regarding the red shift observations carried out at Mount Wilson.

suitable lodgings for the Einsteins, since a real estate agent sent him photos and descriptions of several homes in San Marino and West Pasadena. The rent discussed was exorbitant: $1,500 to $2,000 per month.[39]

By November 12, 1930, Einstein himself confirmed rumors that he would visit Pasadena. So did Millikan, who was ill with a cold. Through his wife Greta Millikan, he replied to inquiries by the *Pasadena Star News* that "invitations had been sent to Dr. Einstein but he had no definite word as to [Einstein's] plans." During the months of November and December 1930, national, international, and local newspapers carried daily reports on Einstein, his travels, and the work of scientists and astronomers at the Mount Wilson Observatory and at Caltech.

When Einstein departed New York on the SS *Belgenland* for Pasadena at midnight on December 14, 1930, journalists attuned to political developments asked whether it was true he might not return if the Fascists took over in Germany: "One should not speak publicly about conditions which he hopes will not come to pass. Still less should one under such circumstances make any decisions in advance or even make public such decisions," he answered. Nevertheless, in Berlin, rumors were circulating that Einstein "may take up his residence in some country other than Germany," such as in a "quiet resort in Southern France where he could carry on his work quietly if the Hitlerites ever gain the upper hand at home."[40]

His activities in New York and California strongly suggest that Einstein would do just that. Upon arrival in New York in December 1930, Einstein engaged for a few days in frenetic social events, such as attending a "Hanukah (fund-raiser) meeting with supposedly 18,000 people in a gigantic hall (Madison Square Gardens) with loudspeakers," he noted in his diary. Apparently Einstein had to be escorted to his hotel by a double row of police guards. He also attended a concert conducted by Arturo Toscanini at the Metropolitan Opera and had "tea with Rockefeller," with a discussion "about a grant organization." Einstein also gave a speech about academic leadership of universities and another on conscientious objection to military service, where he met Helen Keller. Helen Dukas noted in her diary that Einstein's eyes filled with tears on meeting Keller.

We learn much from Einstein's travel diaries, which he kept as an aide memoire for later use by his stepdaughters and family back home.[41] The diary and Einstein's answers to journalists clearly indicate that the seeds of his emigration to America were being sown in New York in 1930. Significantly, the New York engagements also included discussions with Abraham Flexner, whom he met on the third day in New York at the home of Dr. Emanuel Libman, where the subject of their conversation concerned "the organization of scientific institutes."[42] Flexner, together with Veblen and others, was planning an institute devoted exclusively to research—namely, the planning of what would shortly become the Institute for Advanced Study in Princeton.

Einstein arrived in Pasadena for his first visit on New Year's Eve 1930. Shortly after arrival, Elsa Einstein and Helen Dukas, Einstein's assistant, would eventually find a "cute gingerbread house with a shingle roof. Immediately afterwards everything was published with photos and detailed descriptions in the yellow press of Pasadena." The house on S. Oakland Street, about 1 mile west of campus, had a "grand piano, nice kitchen, electric refrigerator, automatic gas heating" for $225 per month. They moved in on January 3, 1931.[43]

We have no evidence that at that time Einstein had further discussions with either Flexner or Veblen concerning the Princeton Institute. But we do know that Fleming and Millikan were eager to have him back the next year. In August 1931, Einstein wrote Millikan that he "thinks with gratitude and pleasure" on his stay in Pasadena "which I hope will not have been my last one. Further thoughts concerning Hubble's observations have proved that the phenomenon adapts itself very well to the theory of relativity."[44]

FIGURE 8.5. Copyright 1931 *Los Angeles Times*. Used with permission.

Arthur Fleming, in private exchanges, made Einstein an extravagant offer that was discussed at a Board of Trustees meeting in Pasadena on September 29, 1931. Fleming had reportedly corresponded with Einstein "with a view to his making a permanent connection with the Institute; that as a result Prof. Einstein had indicated that he would be willing to make such a connection upon the following terms, namely; (1) that he should be paid an honorarium of $5,000 per annum in consideration of his publishing the principal part" of his original scientific work from the California Institute of Technology; (2) that he should be paid an honorarium of $15,000 per annum, inclusive of all expenses, for such years as he should spend not less than ten weeks in residence at the Institute; and (3) that the Institute should pay Mrs. Einstein, if she survives Prof. Einstein, a life annuity of $3,000 per annum beginning at the date of his death." Fleming was intent on securing "the necessary funds and would himself guarantee the $20,000 necessary for the first year."[45] Faculty and trustees were dismayed at Fleming's private negotiations, given Caltech's precarious financial conditions at the time, when no funds were available to build the biology and chemistry laboratories since Caltech's investments had lost significant value during the recession.[46] Millikan traveled to Europe and visited Einstein at his summer home in Caputh during the first week of October. Millikan apparently explained the financial difficulties that Fleming's offer entailed for Caltech, to the extent that Elsa Einstein followed up with a letter in which she wrote that if, in the future, the situation improves, and the invitation still stands, the Einsteins would return to Caltech. She urged Millikan not to make sacrifices.[47]

Two days later, aware that he might have scared Einstein off, Millikan offered an honorarium of $7,000 for future years and insisted that, for 1932, "in view of some previous correspondence which I believe has already been had, I am sure the trustees expect us to decide upon a considerably larger figure and they are prepared to meet it whatever it is that we determine upon. In other words, they want you, and we all want you very much to come. I hope that you will let me regard your letter merely as part of the process of getting the situation clearly before our minds rather than as a definite decision not to visit Pasadena this winter."[48] Some further conversations must have taken place in person during the week of October 14, when both Einstein and Millikan were in Vienna. Einstein spoke on that day at the Physics Institute of the University of Vienna about his collaboration with Walther Mayer on a five-dimensional theory and the rather slim hopes he held for the success of a unified field theory, a project he had been engaged with for a decade.[49]

Einstein was indeed, if only temporarily, upset "with Fleming's way of dealing with proposal and counterproposal." He had expected to have a contract by September 3, 1931, that never arrived. Consequently, he wrote Millikan, he had decided to spend the winter in southern Europe and there gather strength for future negotiations.[50]

But Millikan returned to Berlin, and on November 11, 1931, Einstein was offered a written appointment for the upcoming winter quarter, which he accepted three days later. His duties were "to confer with the members of the staffs and with the graduate students at the Institute and at the Mt. Wilson Observatory as he may wish, to attend such seminars, and to give such lectures as may seem to him desirable during this quarter for a period of from 9 to 12 weeks. It is understood that the compensation which the Institute will make to professor Einstein for this service is to be seven thousand ($7,000.) and that in addition the Institute will provide Professor Einstein and Mrs. Einstein with a secretary during the course of their stay in Pasadena."[51]

On this second visit to Pasadena in 1932, his diary gives clear indication that Einstein then negotiated his upcoming appointment at the Institute for Advanced Study: "Today [22 January 1932] I talked to Veblen, who is here for the moment. Very smart but a bit snobby (discussion on the balcony). I am sitting out there in the sun like a crocodile every day." Ten days later, Veblen visited Einstein again: "we dealt a lot with his theory of electricity. Mathematicians like him are empirical scientists with a pencil and paper. But in general I like him." The next day, they walked together to the astronomy seminar, but they arrived too late. They discussed an appointment and financial support for Walther Mayer, Einstein's collaborator: "Here you only get money to invite those who are already famous. (Like Rockefeller scholarship!)," Veblen told Einstein. "But if you meet him alone, [Veblen] is a nice and unpretentious person. This is so often the case," Einstein noted in the diary. Evidently Einstein was preparing himself to have Veblen as a colleague.

Einstein departed Pasadena in early March, and soon thereafter Millikan initiated arrangements for a third visit. He hoped that Einstein would come under the same conditions, with financial support from the German American Oberlaender Trust, "so that you can feel quite comfortable in view of the fact that the arrangement is thus one which does not make any drain upon the Institute's finances. We are all counting on your being again at the Athenaeum, this time in Mr. Fleming's suite, beginning January 1st [1933]."[52]

Despite Millikan's efforts, Einstein informed him in June 1932 that he eventually planned to go to the new Institute for Advanced Study: he would spend the winter term of 1934 there, but he agreed to come to Pasadena in 1933. He asked that the news be kept secret. He also explained that his assistant Mayer would be able to have a position in Princeton as well, and how important that arrangement was to him. But no official invitation from Caltech arrived during the summer, so that by August, Elsa Einstein rather urgently asked Millikan to decide whether he wanted Einstein at Caltech in 1933, or not. She indicated that, in the future, Einstein might come to Pasadena "ab und zu"—only occasionally, and that Einstein had accepted the Princeton appointment because of the possibilities offered to Mayer.[53] A few days later she also wrote to Veblen, who apparently had visited the Einsteins not long before. Millikan had apparently contacted Abraham Flexner, expressing the idea that Einstein should not "belong to any particular institution; he wants [Einstein] to visit a different institution every year for short periods of time. This is not possible, and since Einstein has given his word to Flexner, he will keep it. Whether we shall visit Pasadena or Princeton in 1933 is unclear, since Einstein is very friendly with Millikan and has great sympathy for him." Einstein himself had left Berlin for Belgium and the Netherlands in order to distance himself from the "unpleasant political events here," she wrote.[54] Following a violent shootout between the SA and Communists in July 1932, Chancellor Franz von Papen had taken major steps toward the implementation of a nationalist government and the destruction of the Weimar Republic by declaring a takeover of the Free State of Prussia and arresting its top government members.

Millikan waited for two months before responding to the Einsteins' inquiries concerning a visit in 1933 and eventually wrote that the agreement with "Princeton will not have any influence whatsoever upon our desire here in Pasadena to go ahead with arrangement made for next winter." He was still "hoping that we shall have the opportunity to discuss matters more fully then and that some arrangements may yet be made between the California Institute of Technology, yourself, and Mr. Flexner, which will enable you to spend some of the time after 1933 here in California, but at present this is merely a hope."[55] Knowing full well the terms of Flexner's offer to Einstein, Millikan made no substantive counterproposal for a long-term association.

By late August 1932, German newspapers were reporting that Einstein would spend five months of the year at the "Flexner Research Institute" in Princeton. The German Minister of Culture and Science inquired of the Prussian Academy whether this was indeed the case.[56] Einstein replied: "I have indeed agreed contractually to spend annually five months at the Flexner institute starting in winter term 1933/34. I have spoken about this matter in detail with the President of the Academy, Herr Prof Planck."[57] Thus, the conversations begun with Oswald Veblen in 1927, and with Flexner in 1930, solidified into a permanent, but still officially

FIGURE 8.6. Millikan, Lemaître, and Einstein, 1933.

only part time, appointment in Princeton. Still, from the analysis of the correspondence there appears to be no doubt that by the middle of 1932, Einstein, Flexner, and Millikan knew that Einstein would not return to Germany.

Einstein arrived for the last time in Pasadena on January 1, 1933. He took long strolls with Georges Lemaître, who was visiting Caltech as well. Journalists noted that they were speaking of the little lambda (λ) and joked about the little lamb that followed them, as Lemaître recalled many years later.

Einstein talked as well with the "young and bright" physicist Robert Oppenheimer, and he spent several evenings with his friend Charlie Chaplin. He prepared for his stay in Princeton by warning Veblen that, "For now, my English is still such that I cannot be held responsible for its effect on the digestive system of the listeners. But this will surely improve."[58]

Ironically, on January 23, 1933, Einstein spoke at the Pasadena Civic Auditorium in a symposium "On German American Agreement." Caltech legal scholar and social scientist William Bennett Munro also spoke on that occasion on the era's economic and political woes. Together with Millikan, Munro was on the board of the Human Betterment Foundation, which had been established in Pasadena five years earlier and had become one of the most engaged eugenic organizations in the world. A majority of its activities, supported by myriad local business people, doctors, and educators, focused on spreading the gospel of forced sterilization.

A week later, on January 30, Hitler came to power in Germany: as in other instances, the most significant events—such as this one—are not mentioned in Einstein's diaries or correspondence. Yet social engagements planned in advance continued unabated. Millikan wrote to Einstein that there would be a dinner at the Athenaeum, "a private affair in the nature of a men's smoker, with quite a group of German-Americans"; there was to be no formal speaking but he hoped to have some informal discussion of German-American problems. "This being the day before your departure, it will serve as an opportunity for some of the many friends you have made here to say 'Auf Wiedersehen.'"[59]

On March 10, an earthquake of 6.4 magnitude struck in Long Beach. Einstein, who was apparently walking on the Caltech campus, reportedly failed to notice.

But before leaving Pasadena for the East Coast the next day, Einstein declared for the press that "As long as I have any choice in the matter, I shall live only in a country where civil liberty, tolerance, and equality

FIGURE 8.7. Long Beach earthquake, March 10, 1933 at 5:54 pm. LAT Staff Photographer. Copyright © 2008. Los Angeles Times. Used with permission.

of all citizens before the law prevail."[60] He spent the summer of 1933 in Belgium and England, and emigrated to the United States in October 1933. He lived in Princeton until his death two decades later, and never visited Pasadena again. But he is still remembered there for the cheerful images he left behind and for the scientific legacy that inspired so many, including those who discovered gravitational waves one hundred years later.

9

How General Relativity Shaped Twentieth-Century Philosophy of Science

DON HOWARD*

A Different Kind of Gravity Wave

The nearly simultaneous publication of the general theory of relativity in the late fall and winter of 1915–1916 by Albert Einstein and David Hilbert was a watershed moment in the history of physics. Even more so than with the introduction of the special theory of relativity ten years earlier, the old, Newtonian view of the universe was dramatically transformed, and the reverberations are felt still today as we struggle to better understand the physics of the big bang and cosmic expansion, black holes, gravitational lensing, dark matter, dark energy, and gravity waves. But it was also a watershed moment in the history of the philosophy of science, one whose effects are also still felt today. Philosophy of science after 1915 is importantly different from what went before. It would be only a mild exaggeration to say that the agenda of twentieth-century philosophy of science was fundamentally reconfigured by the effort to understand a radically new metaphysical picture of the universe, to resolve a variety of deep conceptual questions internal to the general theory of relativity, and to assay the empirical credentials of general relativity.

Much of the fine structure of early twentieth-century philosophy of science, in both the core logical empiricist tradition and the tradition from Ernst Cassier to Hermann Weyl that Tom Ryckman dubs "critical idealism" (Ryckman 2005), was shaped in detail by the challenge posed by general relativity. The same is true for still other traditions and thinkers, including Émile Meyerson (Meyerson 1925; Einstein 1928) and Henri Bergson (Canales 2015). In this chapter, however, I will focus mainly on the development of logical empiricist philosophy of science.

The challenge of general relativity touched on several points, foremost among them being questions about the logical structure and empirical interpretation of physical theory and questions about the ontology of physical theory. In the main, the effect was to structure the philosophical argument space as one pitting varieties of neo-Kantianism against what Moritz Schlick heralded as a new kind of empiricism, one not reducible to the "sensualism" that he attributed to Ernst Mach (Schlick 1921, 323). This new view is the "logical empiricism" famously associated with the Vienna Circle that Schlick, himself, was to lead for thirteen years (Stadler 1997).

As we proceed, it will be well to remember that the logical empiricist tradition was divided into two camps. What we later came to remember as the mainstream logical empiricism of the 1920s and beyond, often today designated as "right-wing" logical empiricism, held that theories are connected with experience and the world via "coordinating" definitions that link individual empirical primitive terms—think of the example of the infinitesimal metric interval—with corresponding definite experiences or structures in the world, such as practically rigid measuring rods and practically regular clocks. This view assumes a principled distinction between analytic coordinating definitions and synthetic empirical propositions, and it confines the moment of convention in science to the choice of those coordinating definitions. On this view, given a choice of coordinating definitions, experience renders a definite verdict on the truth or falsity of each, individual, empirical claim. This way of regarding the empirical interpretation of theories derives from Henri Poincaré's conventionalism about metrical geometry. It was preferred by many because it was thought to offer an especially strong response to Kantian critics of general relativity. This was the position defended mainly by Schlick, Hans Reichenbach, and, in some respects, Rudolf Carnap.

The chief dissent to this view, famously associated in later years with the "left-wing" logical empiricist, Otto Neurath, and the American philosopher, W.V.O. Quine, asserts that theories are tested only as wholes, there being no possible principled distinction between analytic definitions and synthetic empirical propositions, in consequence of which the moment of conventionality is distributed over the entire theory, entailing an in-principle underdetermination of theory choice by experiment. This view derives from the conventionalism of Pierre Duhem (Duhem 1906). Within the Vienna Circle, this left-wing view was also shared by the physicist, Philipp Frank, and the mathematician, Hans Hahn. And this was also Einstein's view. As we shall see, the differences between right-wing and left-wing logical empiricism had major implications for how one understood the empirical support for general relativity.

One might have thought that Einstein was mainly just a passive observer of the developments in philosophy of science engendered by the advent of general relativity. But Einstein was not merely the physicist whose work produced the challenge, he was, himself, one of the most important participants in the ensuing debates, regularly and substantively engaging with all of the major figures. We shall see that, eventually, he came to dissent from both critical idealism and right-wing logical empiricism with a philosophy of science of his own that displays deep affinities with the broadly Duhemian, left-wing variety of logical empiricism.

I will tell the story of general relativity's impact on the philosophy of science by means of two narratives. The first will track questions about the ontological commitments and empirical interpretation of general relativity, highlighting the essential role of two concepts. "Point coincidences," such as the intersections of two world lines, were held by Einstein to constitute the fundamental, invariant, ontological basis of general relativity and the fundamental observables of the theory. "Univocalness," or "*Eindeutigkeit*" in German, names the property of a theory's fixing for itself a univocal or at least structurally univocal representation of the phenomena that it aims to describe. Einstein held that a general methodological requirement on all theory construction was that the theory be univocal.

The second narrative will track debates about neo-Kantian and conventionalist responses to general relativity, culminating in the emergence around 1921 of Schlick's new kind of empiricism, the aforementioned, right-wing, logical empiricism. The two narratives begin in the late fall and winter of 1915–1916, diverge for a time, and then intersect, again, in the early 1920s. Both narratives will pay careful attention to Einstein's central role in the debates. The chapter will conclude with consideration of how and why Einstein finally parted ways with Schlick, Reichenbach, and Carnap, in spite of their many years of fruitful, philosophical exchange concerning all of the core questions at issue.

Two Important Background Facts

Two background facts will prove to be essential for understanding the history that we are exploring.

1. *Einstein the Student of History and Philosophy of Science.* The first of these two facts is that, even by comparison with the higher standards of his day, Einstein was uncommonly well educated in philosophy, generally, and in the history and philosophy of science, more specifically. Einstein's philosophical education began at the very early age of thirteen, when a family friend, Max Talmey, gave him all three of Immanuel Kant's *Critiques* (Kant 1878a, 1878b, 1878c). Talmey later recalled:

> I recommended to him the reading of Kant. At that time he was still a child, only thirteen years old, yet Kant's works, incomprehensible to ordinary mortals, seemed to be clear to him. Kant became Albert's favorite philosopher after he had read through his "Critique of Pure Reason" and the works of other philosophers. (Talmey 1932, 164)

It was especially during his years of study at the Eidgenössische Technische Hochschule (ETH—the Swiss Federal Polytechnic) in Zurich from 1896 to 1900 that Einstein embarked upon what became a lifetime of serious study of history and philosophy of science. In 1897, Einstein's close friend, Michele Besso, first recommended that Einstein read Mach's two books on the history of mechanics and the theory of heat (Mach 1896, 1897). He later recalled this in a letter to Besso:

> As far as Mach's influence on me is concerned, it has certainly been great. I remember quite well that you directed me to his *Mechanik* and *Wärmelehre* sometime during my first years of studies and that both books made a great impression on me. (Einstein to Besso, January 6, 1948, in Speziali 1972, 391)

And he expanded upon this in a 1952 letter to his biographer, Carl Seelig:

> While I was a student, sometime in the year 1897, my attention was drawn to Mach's "*Mechanik in ihre Entwicklung*" by my friend Besso. With its critical attitude toward fundamental concepts and laws, the book made a deep and lasting impression upon me (far more than the *Prinzipien der Wärmelehre*, which I read later). (Einstein to Seelig, April 8, 1952, EA 39-018)

Einstein's correspondence from this period and the surviving contents of his personal library point us to other authors and works that he likely first encountered during his years at the ETH. For example, the library includes Eugen Dühring's *Kritische Geschichte der Principien der Mechanik* (Dühring 1887) and Ferdinand Rosenberger's now little-known but remarkable book, *Isaac Newton und seine physikalischen Prinzipien* (Rosenberger 1895). Dühring was an idiosyncratic thinker, one who mixed democratic socialism, positivism, a quirky form of materialism, anti-Semitism, and the promotion of women's rights with serious work on political economy and the history of science and philosophy. Blind for most of his adult life and disputatious in the extreme, he so enraged his senior colleagues that the University of Berlin withdrew his license to lecture in 1874 (see Cahan 1994; Stuart 1910). Rosenberger's book on Newton is interesting not only for its incisive analysis of virtually all of Newton's scientific work, but also for its inclusion of many long quotations from Newton's correspondence with Henry Oldenburg, the first secretary of the Royal Society, including Newton's letters debating Robert Hooke's critique of Newton's early work on gravity and optics, letters that were otherwise not available in a German translation at that time (see Knott 1907). Newton biographer Richard Westfall wrote of Rosenberger's book that it was "not yet surpassed" (Westfall 1980, 878).

It was also at the ETH that Einstein began the formal, academic study of philosophy, when, in the Summer semester of 1897, he enrolled in August Stadler's lecture course on "Die Philosophie Kants" (*CPAE-1*, Appendix E, 364). Stadler was a prominent figure in the Marburg neo-Kantian community, Hermann Cohen's first PhD student at Marburg, and well known at the time for his work on Kant's teleology and philosophy of matter (Stadler 1874, 1883; see Beller 2000, 85–87; Giovanelli 2003; and Cohen 2015). His Kant course at the ETH provided a comprehensive overview of Kant's life and work and is noteworthy for stressing Kant's engagement with the natural sciences. In the Winter semester of 1897/1898, Einstein enrolled for Stadler's course on the "Theorie des wissenschaftlichen Denkens," which was, as noted in Einstein's ETH transcript, an "Obligatorisches Fach," a required course (*CPAE-1*, Doc. 28, 46). This was mainly a course on deductive logic, but it concluded with a discussion of "Induction and Analogy" (*CPAE-1*, Appendix E, 365–66).

When Einstein moved to Bern in 1902 to take up his new job at the Swiss Federal Patent Office, he continued his self-study of the best contemporary literature in the history and philosophy of science in the context of the informal, weekly discussion group that he organized with his new friends, Conrad Habicht and Maurice Solovine, a group grandiloquently named the "Akademie Olympia." Thanks to Solovine, we have a partial reading list for the Olympia Academy (Solovine 1956, 8–9).

Philosophers of science even today might envy Einstein the opportunity to have read and discussed titles such as these:

- Richard Avenarius. *Kritik der reinen Erfahrung* (Avenarius 1888–1890)
- Richard Dedekind. *Was sind und was sollen die Zahlen?* (Dedekind 1888)
- David Hume. *A Treatise of Human Nature* (Hume 1739)[1]
- Ernst Mach. *Die Analyse der Empfindungen und das Verhältnis des Physischen zum Psychischen* (Mach 1900)
- John Stuart Mill. *A System of Logic*. (Mill 1872)[2]
- Karl Pearson. *The Grammar of Science* (Pearson 1892)[3]
- Henri Poincaré. *La science et l'hypothèse* (Poincaré 1902)[4]

At least two of these readings made a crucial difference, on Einstein's own account, in the genesis of the theory of relativity. In a December 1915 letter to Schlick, in which Einstein complimented Schlick on his new essay, "Die philosophische Bedeutung des Relativitätsprinzips" (Schlick 1915), Einstein wrote:

> You are also quite right with your account of how positivism stands close to the theory of relativity but without requiring it. You have also rightly seen that this direction of thought had been of great influence on my efforts, especially E. Mach and even more Hume, whose *Treatise* I read with eagerness and enthusiasm shortly before the discovery of the relativity theory. (Einstein to Schlick, December 14, 1915 [*CPAE-8*, Doc. 165, 220])

By 1905, Einstein was thus intimately familiar with all of the best contemporary literature on the history and philosophy of science, and that study changed the way in which he did physics. In later years, he reflected upon the benefits of such study, writing to one correspondent:

> I fully agree with you about the significance and educational value of methodology as well as history and philosophy of science. So many people today—and even professional scientists—seem to me like somebody who has seen thousands of trees but has never seen a forest. A knowledge of the historic and philosophical background gives that kind of independence from prejudices of his generation from which most scientists are suffering. This independence created by philosophical insight is—in my opinion—the mark of distinction

> between a mere artisan or specialist and a real seeker after truth. (Einstein to Robert A. Thornton, December 7, 1944, EA 6-574)

These words echo what Einstein had said twenty-eight years earlier in an obituary for Ernst Mach:

> How does it happen that a properly endowed natural scientist comes to concern himself with epistemology? Is there no more valuable work in his specialty? I hear many of my colleagues saying, and I sense it from many more, that they feel this way. I cannot share this sentiment. When I think about the ablest students whom I have encountered in my teaching, that is, those who distinguish themselves by their independence of judgment and not merely their quick-wittedness, I can affirm that they had a vigorous interest in epistemology. They happily began discussions about the goals and methods of science, and they showed unequivocally, through their tenacity in defending their views, that the subject seemed important to them. Indeed, one should not be surprised at this. (Einstein 1916b, 101–2)

In Einstein's view, the chief benefit to the physicist of the study of history and philosophy of science is, precisely, "independence of judgment" or "independence from prejudices of his generation from which most scientists are suffering."

2. *The Philosophers' Engagement with Physics*. The second crucial background fact is that all of the philosophers centrally involved in the construction of what Schlick termed a new form of empiricism capable of responding to neo-Kantian critiques of general relativity were extremely well educated about the relevant physics.

Consider the core figures—Schlick, Reichenbach, and Carnap—starting with Schlick, himself. Most of us rightly remember Schlick, the leader of the Vienna Circle, as a philosopher of science. But his PhD was in physics, with a 1904 Berlin dissertation under the direction of Max Planck, "Über die Reflexion des Lichtes in einer inhomogenen Schicht" (Schlick 1904). After a few years of physics work as an *Assistent* in Göttingen, Schlick retooled as a philosopher through independent study at the University of Zurich, and then he began his formal career in philosophy as a Privatdozent at the University of Rostock in 1911. In 1915, he published his first major paper on relativity, the aforementioned essay, "Die philosophische Bedeutung des Relativitätsprinzips" (Schlick 1915). That was followed by his hugely successful 1917 monograph, *Raum und Zeit in der gegenwärtigen Physik* (Schlick 1917), which went through four German editions by 1922 and was translated into English in 1920 (Schlick 1920). Three years after his magisterial *Allgemeine Erkenntnislehre* (Schlick 1918), he was finally appointed as an *Aussenordentlicher Professor* in Rostock in 1921. The next year he moved to Kiel as *Ordentlicher Professor*, and in 1922 he moved to Vienna, taking up Mach's old chair in the history and philosophy of science, and beginning his famous role as leader of the Vienna Circle until his assassination by a demented former student in 1936. As we will see, the 1915 paper and the 1917 monograph became the basis for a years-long, very close, personal-professional relationship between Einstein and Schlick, one built on the basis of their joint efforts to clarify and reinforce the empirical credentials of the general theory of relativity (see Howard 1984, 1994).

Reichenbach was trained originally as a philosopher, writing his PhD dissertation in 1915 on probability theory under the philosopher, Paul Hensel, and the mathematician, Max Noether, at Erlangen (Reichenbach 1916). But during several years of independent study in Berlin from 1917 to 1920, he audited Einstein's lectures on general relativity and struck up something of a friendship with Einstein. In 1920 he published his first book on relativity, *Relativitätstheorie und Erkenntnis Apriori* (Reichenbach 1920), and that was followed by two more books on relativity in the next eight years, *Axiomatik der relativistischen Raum-Zeit-Lehre* (Reichenbach 1924) and *Philosophie der Raum-Zeit-Lehre* (Reichenbach 1928). In 1926, after serving for six years as a Privatdozent at the Technische Hoschule in Stuttgart, Reichenbach was named an Ausserordentlicher Professor at the University of Berlin, but in the physics, not the philosophy department. The story is that when Einstein wanted to bring Reichenbach to Berlin the philosophers were not interested, so Einstein and Planck instead championed the creation of a chair in philosophy of science in the physics department (see Hecht and Hoffmann 1982; Howard 2014, 361–62). Reichenbach continued to teach at the University of Berlin until, in 1933, he was forced to emigrate, first to Istanbul, Turkey, and then, in 1938 to the United States, teaching at UCLA until his death in 1953. It was while he was working at UCLA that Reichenbach published his last book on the philosophy of physics, *Philosophic Foundations of Quantum Mechanics* (Reichenbach 1944).

Carnap, too, was deeply engaged with physics, especially early in his career. He began his university studies in physics at Jena in 1910, and, after three years of military service he was allowed to resume his studies at the University of Berlin during the 1917–1918 academic year while still serving in the army at a military research institute. It was in the spring of 1918 that Carnap and two friends circulated a "Rundbrief über Relativitätstheorie," which includes several comments by Carnap that indicate his growing interest in foundational and philosophical questions (Carnap 1963, 10). Back at Jena after the First World War, Carnap completed a PhD dissertation on the problem of space under the supervision of the neo-Kantian philosopher, Bruno Bauch (Carnap 1921). This is a careful and nuanced discussion of the impact of general relativity on our understanding of space, one of Carnap's core arguments being that we now see that only topological, not metrical structure can be regarded as an a priori feature of general relativistic space-time, much as Cassirer argued in his book on Einstein's theory of relativity in that same year (Cassirer 1921).

Over the next few years, much of Carnap's work dealt with the philosophy of physics in one way or another, as with a very nice 1925 paper on the interdependence of spatial and temporal properties (Carnap 1925) or his too-little-known 1923 essay, "Über die Aufgabe der Physik" (Carnap 1923) and his equally neglected 1926 monograph on *Physikalische Begriffsbildung* (Carnap 1926), both of which show him moving steadily away from Kant in the direction of conventionalism (see Richardson 1992). After his *Habilitation* in Vienna in 1926 with the thesis that would become his famous and influential *Der logische Aufbau der Welt* (Carnap 1928) and his move to a full professorship in Prague in 1931, Carnap's interests shifted more toward general philosophy of science and philosophy of language. But his interest in the philosophy of physics remained strong, as witness the posthumously published volume, *Two Essays on Entropy* (Carnap 1977).

Similar stories can be told about a number of other figures, such as Cassirer, who, though lacking formal training in physics, evinced a sophisticated understanding of very technical material in his just-mentioned book on the theory of relativity or his somewhat less well-known book on the conceptual foundations of quantum mechanics (Cassirer 1937). An especially interesting example is the physicist and philosopher of science, Philipp Frank, who was Einstein's successor at the Charles University in Prague in 1912—highly recommended by Einstein himself—and later, after his emigration to the United States in 1938, where he secured an appointment in the physics department at Harvard, Einstein's biographer (Frank 1949a). A prominent member of the Vienna Circle, Frank wrote often on issues in the foundations of physics, as with his 1932 book on the law of causality and its limits (Frank 1932), or his 1938 monograph on *Interpretations and Misinterpretations of Modern Physics*

(Frank 1938). Frank was especially interested in the work of Mach and Duhem (see Frank 1917), a recurring theme being the manner in which, on Frank's analysis, the philosophies of science of Einstein and Duhem effected a merger and integration of Mach's positivism and Poincaré's conventionalism (see Frank 1949b; Howard forthcoming). Frank did not play a central role in the debates about general relativity and the philosophy of science that we will now begin to track. But his example nonetheless illustrates the many deep connections between physics and philosophy of science in the first half of the twentieth century (see Howard 2004).

On the Path to General Relativity: The "Hole" Argument

Our main story begins with Einstein's struggle from early 1913 to late 1915 to find the correct formulation for a general theory of relativity, a struggle that culminated with the publication in November 1915 of three papers that lay out the essentials of the modern, general theory of relativity (Einstein 1915a, 1915b, 1915c) and his magisterial review paper of early 1916, "Die Grundlage der allgemeinen Relativitätstheorie" (Einstein 1916a). An earlier version of the theory, now commonly referred to as the "*Entwurf*" theory, after the first word in the title of the paper introducing it (Einstein and Grossmann 1913; see also Einstein and Grossmann 1914 and Einstein 1914), differs importantly from the final version of the theory in that it has restricted covariance properties. Einstein had convinced himself that a fully generally covariant theory of gravitation would not work for several reasons. For example, he thought, wrongly, that such a theory would not yield the correct Newtonian limit. More important for our purposes, however, is another reason. Einstein had also thought, wrongly, that a fully covariant theory would fail to satisfy the aforementioned general methodological requirement that he termed "Eindeutigkeit" or "univocalness," according to which a theory must fix for itself a unique solution, given a specified set of boundary conditions.

> *Events in the gravitational field cannot be determined univocally by means of generally covariant differential equations for the gravitational field.*
>
> If we demand, therefore, that the course of events in the gravitational field be completely determined by means of the laws that are to be established, then we are obliged to restrict the choice of the coordinate system. (Einstein 1914, 1067)

Einstein's reasoning was as follows. Consider a region of space-time devoid of matter and energy, a "hole," and a solution of a set of generally covariant field equations, $G(x)$ covering the hole. Transform this solution to a new set of coordinates, x', yielding $G'(x')$. Consider a specific point, x', y', z', t'. Take the value of G' assigned to that point, $G'(x', y', z', t')$ and assign it now to the point with the numerically identical coordinates in the unprimed coordinates. Do this for all points in the "hole." Precisely because of the general covariance of the field equations, the result is also a solution, $G'(x)$, that, in general, differs from $G(x)$ inside the hole, because one and the same point, x, y, z, t is now assigned two different values, $G(x)$ and $G'(x,)$ while $G'(x)$ comes smoothly to agree with $G(x)$ outside of the hole. But that means that such generally covariant field equations fail to satisfy the "Eindeutigkeit" requirement.

The "hole" argument was a seemingly fatal objection to generally covariant gravitational field equations. It took many months of reflection, analysis, debate, and discussion for Einstein to realize that a serious mistake lurked within the hole argument. When, in November 1915, Einstein published the fully covariant version of the theory, without directly addressing such concerns, more than one correspondent expressed puzzlement about what had happened to the hole argument. This is the moment when the story of general relativity's influence on the development of twentieth-century philosophy of science really begins.[5]

Narrative One: Point Coincidences, "*Eindeutigkeit*," and Both the Ontological Commitments and Empirical Interpretation of General Relativity

In the late summer or early fall of 1915, Einstein began to realize that something was seriously wrong with the *Entwurf* theory. Among other things, he discovered that it gave the wrong prediction for the precession of the perihelion of Mercury. He also diagnosed his error regarding the Newtonian limit in a fully covariant theory of gravitation. Thus began his triumphal march to the final, generally covariant version of the theory. But there remained the problem of the hole argument. Did it not prove, on very general grounds, that a fully covariant theory would be inadmissible? How could there possibly be a mistake in this argument?

Important clues to the solution of Einstein's puzzle about the hole argument were provided by the Göttingen physicist, Paul Hertz, in the wake of Einstein's week-long visit to Göttingen in July 1915 to discuss the gravitation problem with participants in David Hilbert's mathematical physics seminar. Hertz might well have been the first to suggest to Einstein that his mistake was assuming that a coordinate chart, alone, suffices for individuating space-time manifold points. Instead, Hertz seems to have argued, one needs a covariant conception of point identity (see Howard and Norton 1994). Whether Hertz was the source of this idea or not, that is precisely how Einstein, himself, explained the matter to surprised and confused colleagues.

One of those who pressed Einstein about the fate of the hole argument was his close friend, Paul Ehrenfest, to whom he wrote the following on December 26, 1915:

> In § 12 of my work of last year, everything is correct (in the first three paragraphs) up to that which is printed with emphasis at the end of the third paragraph. From the fact that the two systems $G(x)$ and $G'(x)$, referred to the same reference system [the same x], satisfy the conditions of the grav. field, no contradiction follows with the univocalness of events. That which was apparently compelling in these reflections founders immediately, if one considers that (1) the reference system signifies nothing real, (2) that the (simultaneous) realization of two different g- systems (or better, two different fields) in the same region of the continuum is impossible according to the nature of the theory.
>
> In place of § 12, the following reflections must appear. The physically real in the world of events (in contrast to that which is dependent upon the choice of a reference system) consists *in spatio-temporal coincidences.** Real are, e.g., the intersections of two different world lines, or the statement that they *do not* intersect. Those statements which refer to the physically real therefore do not founder on any univocal coordinate transformation. If two systems of the $g_{\mu\nu}$ (or in general the variables employed in the description of the world) are so created that one can obtain the second from the first through mere space-time transformation, then they are completely equivalent. For they have all spatio-temporal point coincidences in common, i.e., everything that is observable. [These reflections show at the same time how natural the demand for general covariance is.] (Einstein to Ehrenfest, December 26, 1915, *CPAE-8*, Doc. 173, 228–29)
>
> * and in nothing else!

Eight days later he gave much the same account to his long-time, intimate friend, Michele Besso:

> In the *Lochbetrachtung*, everything was correct up to the final conclusion. There is no physical content in the existence of two different solutions $G(x)$ and $G'(x)$ with reference to the *same* coordinate system K. Attributing two different solutions to the same manifold is senseless, and the system K has, indeed, no physical reality. The following consideration takes the place of the *Lochbetrachtung*. From a physical point of view, nothing is *real* except the totality of spatiotemporal point coincidences. If, e.g., physical processes were to be built up solely out of the movements of material points, then the meetings of the points, i.e., the points of intersection of their world lines, would be the only reality, i.e., observable in principle. These points of intersection are naturally preserved under all transformations (and no new ones are added), if only certain univocalness conditions are maintained. Thus, it is most natural to demand of the laws that they do not determine *more* than the totality of the spatiotemporal coincidences. According to what has been said, this is already achieved by generally-covariant equations. (Einstein to Besso, January 3, 1916, *CPAE-8*, Doc. 178, 235)

The argument that Einstein presented to Ehrenfest and Besso to explain the mistake in the hole argument is now commonly referred to as the "point coincidence" argument. The key insight is that space-time manifold points must, themselves, be individuated in a covariant manner as points of intersection or coincidence of two world lines. The world line structure is an invariant feature of a generally covariant theory, because the fact that two world lines intersect is true in all frames of reference, whereas the location of that intersection depends on one's choice of a set of space-time coordinates. In this sense, a coordinate chart—a "reference system," as Einstein calls it—signifies nothing real. Reality consists only in invariant structural features of the theory, at base the point coincidences. Though readers could easily have missed the point, Einstein had already said precisely that in the last of his three general relativity papers from November 1915, where he wrote: "The relativity postulate in its most general formulation, which turns the space-time coordinates into physically meaningless parameters, leads with compelling necessity to a completely determinate theory of gravitation" (Einstein 1915c, 847).

With hindsight, looking backward through the lens of modern differential geometry, the essence of the point coincidence argument seems almost trivial. But that was not the case in the winter of 1915–1916. At the time, thinkers like Einstein routinely conflated frames of reference with systems of coordinates and did not understand the difference between passive and active transformations. It was, in fact, a hard-won insight.

Note that the point coincidence argument appeals to the very same methodological principle—the Eindeutigkeit or univocalness principle—that was the foundation for the hole argument. Only now Einstein argues that, if we individuate manifold points in an invariant manner, as points of intersection between two world lines, then a fully generally covariant gravitation theory does still fix univocal solutions for itself everywhere in space-time, including the "holes." Eindeutigkeit is saved by our getting clear about the theory's fundamental, point-event ontology.

Given the prominence of the Eindeutigkeit principle in Einstein's thinking about general relativity, one might wonder about the source for this idea. It goes back to an 1895 essay, "Das Gesetz der Eindeutigkeit" (Petzoldt 1895), by a once prominent but, sadly, long since forgotten philosopher of science, Joseph Petzoldt. Petzoldt was a positivist in the tradition of Mach and Richard Avenarius. He earned a PhD in Göttingen in 1890, after which he taught for many years at a Gymnasium in Berlin-Spandau. In 1904, he became a Privatdozent at the Technical University in Charlottenburg, being promoted to Ausserordentlicher Professor in 1922. Petzoldt was the founder of the Gesellschaft für positivistische Philosophie and the *Zeitschrift für positivistische Philosophie* in 1912, as well as the Internationale Gesellschaft für empirische Philosophie in Berlin in 1927, a group that counted Reichenbach among its members. He was best known for advocating a program of "relativistic positivism," a blend of Mach and relativity theory (Petzoldt 1906, 1912b; see also Dubislav 1929).

In the mentioned 1895 essay, Petzoldt introduced the core idea as a very general principle or postulate:

> We must . . . bring to bear on nature a certain general presupposition, without whose confirmation we ourselves could not live, either mentally or bodily. Such a presupposition lies at the base of all scientific research, something of which we may be more or less consciously aware, and we may be of the firm conviction that it will hold up everywhere, since we could not conceive of ourselves, with our particular mental nature, if we once imagine it being given up. Both our individual constitution and that postulate, as we may designate the relevant presupposition, belong inseparably together. The latter consists in nothing other than the assumption of the *thoroughgoing complete*

> *determination,* or—as we want to say in order to emphasize the most important side of the matter—in the assumption of the *uniqueness of all processes* [*Eindeutigkeit aller Vorgänge*]. (Petzoldt 1895, 167–68)

Notice that Petzoldt's expression, "Eindeutigkeit aller Vorgänge," is very similar to the expression used by Einstein in his letter to Ehrenfest about the hole and point coincidence arguments, "Eindeutigkeit des Geschehens." But that Petzoldt was also the proximal source for Einstein's invocation of the Eindeutigkeit principle in the context of the hole argument in late 1912 or early 1913 is strongly suggested by Petzoldt's own application of the principle to relativity theory in a 1912 paper:

> The task of physics becomes, thereby, the *eindeutige* general representation of events from different standpoints moving relative to one another with constant velocities, and the *eindeutige* setting-into-relationship of these representations. Every such representation of whatever totality of events must be *eindeutig* mappable onto every other one of these representations of the same[1)] events. The theory of relativity is one such mapping theory. What is essential is that *eindeutige* connection. Physical concepts must be bent to fit for its sake. We have theoretical and technical command only of that which is represented *eindeutig* by means of concepts.
>
> [[1)]Better: representations of events in arbitrarily many of those systems of reference that are *eindeutig* mappable onto one another are representations of "*the same*" event. Identity must be *defined,* since it is not given from the outset. (Petzoldt 1912a, 1059)]

Einstein and Petzoldt had to have been in contact by 1912, when Petzoldt recruited Einstein as one of the signatories for his call to establish the new Gesellschaft für positivistische Philosophie. From shortly after Einstein's move to Berlin, he was in regular contact with Petzoldt, who seems to have attended Einstein's lectures on relativity at the University of Berlin in 1914–1915 (see Howard 1991). And we glean from their correspondence that Einstein had a generally high regard for Petzoldt's work. In a letter dated April 14, 1914, Einstein writes about Petzoldt's essay, "Die Relativitätstheorie der Physik" (Petzoldt 1914):

> I read your comments on relativity theory in the Zeitschr. für posit. Philosophie with much pleasure. From it I see with astonishment that you are closer to me in your understanding of the subject, as well as

> with regard to the sources from which you draw your scientific convictions, than my true colleagues in the field, even as far as they are unconditional supporters of relativity theory. (Einstein to Petzoldt, April 14, 1914, *CPAE-8*, Doc. 5, 16)

There is a long history of the wider reception of Petzoldt's advocacy of the Eindeutigkeit principle. His original, 1895 paper was cited approvingly by many thinkers whose roles in the development of twentieth-century philosophy of science we are tracking, from Mach to Carnap. And philosophical discussion of the Eindeutigkeit principle in the philosophy of science became entangled for obvious reasons with the concept of categoricity in the foundations of mathematics, which was introduced by David Hilbert (Hilbert 1899) and Oswald Veblen (Veblen 1904) only a few years after Petzoldt's 1895 paper. A formalized mathematical theory is said to be categorical or monomorphic if it fixes the class of its models up to the point of isomorphism. In other words, such a theory yields a structurally univocal description of its intended domain, so it makes sense that Eindeutigkeit and categoricity were often taken to be expressions of the same core desideratum (see Howard 1991, 1996).

Just as Einstein started explaining to Ehrenfest, Besso, and others why the hole argument was fallacious and how Eindeutigkeit in a fully covariant gravitation theory is restored by means of the point coincidence argument, he received from Schlick a copy of Schlick's essay on the philosophical significance of the relativity principle (Schlick 1915). Einstein was impressed:

> Yesterday I received your essay and I have already studied it through completely. It is among the best that have until now been written about relativity. From the philosophical side, nothing at all appears to have been written on the subject that is nearly so clear. At the same time, you really have complete command of the subject. There is nothing in your exposition with which I find fault. (Einstein to Schlick, December 14, 1915, *CPAE-8*, Doc. 165, 220)

Schlick's essay was technically sophisticated and thick with deep philosophical insights, concerning such issues as conventionalism, simplicity as a criterion of theory choice, realism, and the ontological import of physical theories (see Howard 1984). Schlick had discussed general relativity only glancingly, and then only in the form of the early *Entwurf* theory, so Einstein brought him up to speed on the newest developments:

> Your comments on the general theory of relativity are also entirely correct, as far as this theory was right by then. The new finding is the result that a theory exists that agrees with all previous experience whose equations are covariant with arbitrary transformations in the space-time variables. Thus time & space lose the last vestiges of physical reality. (Einstein to Schlick, December 14, 1915, *CPAE-8*, Doc. 165, 220)

"Thus time & space lose the last vestiges of physical reality." With that, Einstein introduced a way of characterizing the chief implication of the point coincidence argument that resonated in the philosophy of space-time literature for a long time.

In his 1916 review article on the foundations of the general theory of relativity, Einstein foregrounded this theme:

> That this requirement of general co-variance, which takes away from space and time the last remnant of physical objectivity, is a natural one, will be seen from the following reflexion. All our space-time ascertainments invariably amount to a determination of space-time coincidences. If, for example, events consisted merely in the motions of material points, then ultimately nothing would be observable but the meetings of two or more of these points. Moreover, the results of our measurings are nothing but verifications of such meetings of the material points of our measuring instruments with other material points, coincidences between the hands of a clock and points on the clock dial, and observed point-events happening at the same place at the same time.
>
> The introduction of a system of reference serves no other purpose than to facilitate the description of the totality of such coincidences. . . . As all our physical experiences can be ultimately reduced to such coincidences, there is no immediate reason for preferring certain systems of co-ordinates to others, that is to say, we arrive at the requirement of general covariance. (Einstein 1916a, 776–77)

All of this must have had a powerful and immediate impact on Schlick who, previously unaware of these latest developments in general relativity, had nonetheless been thinking about closely related issues.

Two of those issues loomed large in Schlick's 1915 paper on the philosophical significance of the relativity principle. The first was Schlick's elaboration of an argument for the underdetermination of theory choice by evidence that he had first given in his 1910 paper on "The Essence of Truth According to Modern Logic" (Schlick 1910) where he introduced a definition of truth as a one-way univocal ("eindeutige") coordination between theory and world. On this view, an individual true proposition, and, by extension, a true theory, consisting of many propositions, is correlated univocally with a uniquely specified state of affairs in the part of the world it aims to describe. But the correlation will generally not be univocal in the other direction, there being, typically, a multiplicity of different propositions and theories that can equally well be univocally correlated with the same state of affairs. Here is how Schlick put the idea in the 1915 paper:

> The totality of our scientific propositions, in word and formula, is in fact nothing else but a system of symbols correlated to the facts of reality; and that is equally certain, whether we declare reality to be a transcendent being or merely the totality and interconnection of the immediately "given." The system of symbols is called "true," however, if the correlation is completely univocal. Certain features of this symbol system are left to our arbitrary choice; we can select them in this way or that without damaging the univocal character of the correlation. It is therefore no contradiction, but lies, rather, in the nature of the matter, that under certain circumstances, several theories may be true at the same time, in that they achieve indeed a different, but each for itself completely univocal designation of the facts. (Schlick 1915, 149)

Whether Schlick's version of conventionalism stands closer to that of Duhem or Poincaré requires a close and careful analysis and will emerge as a crucial question below. For now, appreciate Schlick's characteristic emphasis upon the relationship's being one between "the facts of reality" and the "*totality* of our scientific propositions," which, I maintain, places him closer to Duhem (see Howard 1994).

The second issue is closely related to the first. If multiple theories are equally univocally coordinated with reality, how do we choose among them. Comparative simplicity is the obvious answer, but what makes one theory simpler than another and how can our choosing in this way be justified? Schlick mocks those who say that we choose the simpler theory because the world is simple, for the obvious reason that we know the world only through our theories, meaning that this answer is circular. What was Schlick's answer?

> In very many cases, the greater simplicity of a theory rests upon its containing fewer arbitrary elements. . . . But it is clear now that the greater the number of arbitrary elements that a theory contains, the more in it results from my willful choice, the less from what the facts compel. But naturally we must say that a theory represents reality only to the extent that it is determined just by the objective facts. . . . Naturally, however, we want to exclude from our theories, as far as possible, not only the false but also the superfluous accessories, our own addition. We do this by choosing those with a minimum of arbitrary assumptions, that is, the simplest. Then we are certain to stray at least no farther from reality than required by the limits of our knowledge. (Schlick 1915, 154–55)

Think about how Einstein reacted to these words in December 1915, having just won his way through to the final, fully covariant form of the general theory of relativity, in which process he had to figure out that coordinate charts are superfluous accessories, our own additions, that can therefore represent nothing real, and that only invariant structural features of space-time, such as the point coincidences, are real.

Einstein's relationship with Schlick warmed steadily over the next few years. Einstein became a great champion of Schlick's career (see Howard 1984). Stimulated by their exchange over his 1915 paper, Schlick published in 1917 what became the most successful book yet published on relativity theory, his monograph on *Raum und Zeit in der gegenwärtigen Physik* (Schlick 1917). Schlick sent Einstein a copy of the manuscript early in 1917, describing the main point of the book in this way: "The essay is less a presentation of the general theory of relativity itself and more a thoroughgoing exposition of the proposition that, in physics, space and time have now forfeited all objectivity" (Schlick to Einstein, February 4, 1917, *CPAE-8*, Doc. 296, 388), meaning that the main goal of the monograph was to elucidate the core philosophical lessons of Einstein's point coincidence argument. Einstein was very impressed, writing to Schlick in May: "Again and again I take a look at your little book and am delighted by the splendidly clear expositions. And the last section, 'Relations to Philosophy,' appears to me to be excellent" (Einstein to Schlick, May 21, 1917, *CPAE-8*, Doc. 343, 456).

Much of that last section that Einstein praised so highly is devoted to critiques of Kant and Mach. In this context, a key question for Schlick concerns the emergence of the concept of physical space out of intuitive space or the multiple intuitional spaces corresponding to different sense modalities. After all, the spatiality of tactile experience is "toto genere different" from that of optical experience. "On the other hand, the space of the physicist, which, as objective space, we contrast with those subjective spaces, is only one and is thought of independently of our sense experiences" (Schlick 1917, 53). The answer to the question of the genesis of the concept of objective, physical space is afforded, says Schlick, by Einstein's notion of point coincidences:

> It is important now to make clear which particular experiences lead us to coordinate a specific element of optical space with a specific element of haptic space and, thereby, to construct the concept of the "point" in objective space. It is, namely, experiences of coincidences that are to be taken into consideration here. In order to determine a point in space, one must somehow, directly or indirectly, *point* to it, one must make the point of a compass, or a finger, or a set of cross-hairs, coincide with it, that is, one sets up a spatiotemporal coincidence of two otherwise separated elements. Now it turns out that these coincidences always appear to agree for all the intuitional spaces of the various senses and for various individuals; for just that reason, an objective "point" is defined by them, i.e., one independent of individual experiences and valid for all. . . . Upon more careful reflection, one easily finds that we arrive at the construction of physical space and time exclusively by this method of coincidences and in no other way. The space-time manifold is precisely nothing other than the totality of objective elements defined by this method.
>
> This is the result of the psychological-empiriocritical analysis of the space and time concept, and we see that we encounter precisely *the* significance of space and time that Einstein recognized as alone essential for physics and there gave proper expression. For he repudiated the Newtonian conception . . . and instead founded physics on the concept of the coincidence of events. Here, therefore, physical theory and epistemology join hands in a beautiful alliance. (Schlick 1917, 57–58)

But this is a genetic story about the construction of the concept of objective, physical space. It would be a mistake to infer that, thanks to this construction, physical space is reducible to intuitional space. This becomes the central thesis in Schlick's critique of Mach.

> None of the quantities that appear in physical laws denote "elements" in Mach's sense; the coincidences that are expressed by the differential equations of physics are not capable of being immediately experienced, they do not directly signify a coinciding of sense data, but rather, for the moment, a coinciding of non-intuitive quantities, like electrical and magnetic field strengths and the like. Now nothing compels the assertion that only the intuitional elements—the colors, the tones, etc.—exist in the world; one can equally well assume that, aside from them, there are elements or qualities that are not capable of being directly experienced but that are likewise to be designated as "real," whether they are comparable with those intuitive qualities or not. Electrical forces, for example, can then equally well designate elements of reality as colors and tones. They are indeed *measurable*, and there is no discernible reason why epistemology should repudiate the reality criterion of physics. (Schlick 1917, 59)

Schlick might have been venturing here a gentle critique of Einstein, as well. Recall that, in his December 26, 1915 letter to Ehrenfest, Einstein put the emphasis on the reality of point coincidences: "The physically real in the world of events . . . consists *in spatio-temporal coincidences*." But in his 1916 review article, Einstein emphasized the observability of point coincidences:

> All our space-time ascertainments invariably amount to a determination of space-time coincidences. If, for example, events consisted merely in the motions of material points, then ultimately nothing would be observable but the meetings of two or more of these points. Moreover, the results of our measurings are nothing but verifications of such meetings of the material points of our measuring instruments with other material points, coincidences between the hands of a clock and points on the clock dial, and observed point-events happening at the same place at the same time. (Einstein 1916a, 776)

Are point coincidences taken to be fundamental because of their reality or their observability? What takes priority, ontology or epistemology? Is there a deep connection between the reality and the observability of point coincidences? Are they real because they are observable or are they observable because they are real? An astute reader like Schlick might well have been puzzled enough to want to push Einstein to be clearer about these questions.

Did Einstein take Schlick to be hinting at a confusion in his thinking? There is no direct evidence showing that he did. But there is intriguing indirect evidence. For it was this very argument of Schlick's that drew the only seriously critical response from Einstein, who wrote:

> The second point to which I want to refer concerns the reality concept. Your view stands opposed to Mach's according to the following schema:
>
> Mach: Only impressions are real.
>
> Schlick: Impressions and events (of a phys[ical] nature) are real.
>
> Now it appears to me that the word "real" is taken in different senses, according to whether impressions or events, that is to say, states of affairs in the physical sense, are spoken of.
>
> If two different peoples pursue physics independently of one another, they will create systems that certainly agree as regards the impressions ("elements" in Mach's sense). The mental constructions that the two devise for connecting these "elements" can be vastly different. And the two constructions need not agree as regards the "events"; for these surely belong to the conceptual constructions. Certainly only the "elements," but not the "events," are real in the sense of being "given unavoidably in experience."
>
> But if we designate as "real" that which we arrange in the space-time schema, as you have done in the theory of knowledge, then without doubt the "events," above all, are real.
>
> Now what we designate as "real" in physics is, no doubt, the "spatio-temporally-arranged," not the "immediately-given." The immediately-given can be illusion, the spatio-temporally-arranged can be a sterile concept that does not contribute to illuminating the connections between the immediately given. *I would like to recommend a clean conceptual distinction here.* (Einstein to Schlick, May 21, 1917, *CPAE-8*, Doc. 343, 456–57)

Schlick had, of course, written that the Machian elements of sensation and physical quantities are equally real. Einstein says, not quite, because there is an essential underdetermination in the physical ontology that is not to be found in different people's immediate experiencings of the world, which latter should have come as no to surprise to Schlick who, again in the 1917 monograph, repeated the argument for the underdetermination of theory choice based on his definition of truth as one-way univocal coordination between theory and world.

Einstein's response to Schlick does not resolve the possible confusion about whether point coincidences are privileged epistemologically, ontologically, or both. But the community continued to think about this question for more than a decade. Reichenbach got the last word with a lengthy discussion of the problem in his 1928 *Philosophie der Raum-Zeit-Lehre*. Let us recall that Reichenbach was Einstein's former student, friend, and, since 1926, colleague in the physics department at the University of Berlin, meaning that they had ample opportunity to discuss these matters without there being a paper record of those discussions.

The immediate context is Reichenbach's argument that local topological structure fixes the event structure in space-time and, thus, also the causal structure. In what is clearly a criticism of Schlick, Reichenbach notes:

> Other attempts have been made to explain the topology of space and time. The coordinate system assigns to the system of coincidences, of point-events, a mutual order that is independent of any metric. This order of coincidences must therefore be understood as an ultimate fact. The attempt has been made to justify this order as necessary; it has been regarded as a function of the human perceptual apparatus rather than of the objective world. Accordingly, it has been claimed that sense perceptions supply directly only coincidences, and that the ultimate element of space-time order is determined by the character of our sense perceptions. In this connection appeal is made to the experimental methods of the physicist, in which coincidences of dials and pointers play an important role. (Reichenbach 1928, 327)

He goes on to explain that this is a serious error:

> This view is untenable. First of all, it is obvious that we cannot regard the order of coincidences as immediately given, since the subjective order of perceptions does not necessarily correspond to the objective order of external events. It can serve only as the basis of a complicated procedure by which the objective order is inferred. . . . We must therefore introduce rules for the construction of the objective order; such rules have been formulated in the topological coordinative definitions.
>
> It is a serious mistake to identify a coincidence, in the sense of a point-event of space-time order, with a coincidence in the sense of

> a sense experience. The latter is *subjective coincidence,* in which sense perceptions are blended. . . . The former, on the other hand, is *objective* coincidence, in which physical things, such as atoms, billiard balls or light rays collide and which can take place even when no observer is present. The space-time order deals only with objective coincidences, and we go outside the realm of its problems in asking how the system of objective coincidences is related to the corresponding subjective system. The analysis of this question belongs to that part of epistemology that explains the connection between objective reality, on the one hand, and consciousness and perception on the other. Let us say here only that any statement about objective coincidences has the same epistemological status as any other statement concerning a physical fact.
>
> It is therefore not possible to reduce the topology of space and time to subjective grounds springing from the nature of the observer. On the contrary, we must specify the principles according to which an objective coincidence is to be ascertained. This means that we must indicate a method how to decide whether a physical event is to be considered as one, or as two or more separate point events. (Reichenbach 1928, 327–28)

Reichenbach concludes:

> Objective coincidences are therefore physical events like any others; their occurrence can be confirmed only within the context of theoretical investigation. Since all happenings have until now been reducible to objective coincidences, we must consider it the most general empirical fact that the physical world is a system of coincidences. It is this fact on which all spatio-temporal order is based, even in the most complicated gravitational fields. What kind of physical occurrences are coincidences, however, is not uniquely determined by empirical evidence, but depends again on the totality of our theoretical knowledge. (Reichenbach 1928, 328–29)

With that last remark, Reichenbach seems to endorse the point that Einstein made to Schlick eleven years earlier, with his claim that, while the "two different peoples" are fated to agree on the subjective elements of sensation, they need not and typically would not agree on the space-time event ontology.

Why is it important to get clear about the distinction between what I once termed "pointer coincidences" and "point coincidences" (Howard 1999)? We will shortly take up the promised second narrative about general relativity's impact on the development of twentieth-century philosophy of science. That narrative will foreground questions about the empirical interpretation of general relativity and steps that led to the emergence of mature, verificationist, right-wing logical empiricism of the kind endorsed by Schlick and Reichenbach in the early 1920s. The issue of Eindeutigkeit will reappear.

Narrative Two: Neo-Kantianism, Conventionalism, and the "New Empiricism"

Kant, as well as Mach, was a target of criticism in Schlick's 1917 monograph, *Raum und Zeit in der gegenwärtigen Physik*. Kant was famous for arguing that space and time are necessary a priori forms of outer and inner intuition, respectively, "intuition" being Kant's term for our direct experience either of things as they seem to present themselves to us in the outside world, thus "outer intuition," or as part of our inner experience, "inner intuition." More specifically, Kant had argued that our experience of objects in space must, as a matter of transcendental necessity, be organized under a Euclidean geometrical form. General relativity's positing a physical space-time with variable curvature was a strong, prima facie challenge to Kant's doctrine of space and time.

Schlick's main argument against Kant was that even if one could make clear sense out of Kant's doctrine of space as the form of outer intuition, since physical space is something apprehended in a purely conceptual manner, possible facts about spatial intuition could have no bearing on our theoretical accounts of physical space. In other words, Schlick is arguing that nothing about the psychology of our experience of space can constrain the physicist's investigations.

Perhaps it was Schlick's critical discussion of Kant and relativity that led Einstein to revisit him. Perhaps it was conversations with Reichenbach, who was auditing Einstein's general relativity lectures in the 1917–1918 academic year and would go on to publish his book on *Relativitästheorie und Erkenntnis Apriori* in 1920. Whatever the impetus, Einstein was doing his homework. In the summer of 1918, he writes to Max Born:

> I am reading Kant's *Prolegomena* here, among other things, and am beginning to comprehend the enormous suggestive power that

> emanated from the fellow and still does. Once you concede to him merely the existence of synthetic a priori judgments, you are trapped. I have to water down the "a priori" to "conventional," so as not to have to contradict him, but even then the details do not fit. Anyway it is very nice to read, even if it is not as good as his predecessor Hume's work. Hume also had a far sounder instinct. (Einstein to Born, after June 29, 1918, *CPAE-8*, Doc. 575, 818)

It is noteworthy that Einstein here introduces a theme—recasting the a priori as the conventional–that will become a leitmotif not only in his own work over the next few years but in that of many of his interlocutors, as well.

It was well that Einstein was working out his thoughts about Kant, because, soon after the end of World War I, a veritable flood of new books and papers on Kant and relativity theory began to appear. One of the first that drew Einstein's attention was Ewald Sellien's book on the epistemological significance of the theory of relativity, *Die erkenntnistheoretische Bedeutung der Relativitätstheorie* (Sellien 1919). Einstein was not impressed. He wrote to Schlick in October 1919: "Philosophers are already busily trying to squeeze the general theory of relativity into the Kantian system. Have you seen the quite foolish dissertation by Sellien (a pupil of Riehl)?" (Einstein to Schlick, October 17, 1919, *CPAE-9*, Doc. 142, 204A). As indicated, Sellien was a student of Alois Riehl, the prominent neo-Kantian of the critical realist variety who held the philosophy chair in Berlin from 1905 to his death in 1924 (see von Kloeden 2003). Another one of Riehl's students was Ilse Schneider, who met a number of times with Einstein to discuss her work on Kant and relativity, work that eventuated in one of the better books on the topic (Schneider 1921).[6] Perhaps she suggested to Einstein that he look at Sellien's book, for he shared his reservations with her in September 1919: "I have received the mentioned dissertation by S. [Ewald Sellien] (Epistemology and Relativity Theory). . . . Kant's celebrated view on time reminds me of Andersen's tale about the emperor's new clothes, except that instead of the emperor's clothes, it concerns the form of intuition" (Einstein to Schneider, September 15, 1919, *CPAE-9*, Doc. 104, 155–56).

The rising flood of neo-Kantian critiques of and responses to general relativity was making it clear to Einstein and colleagues like Schlick that the defenders of general relativity had to do a still better job of explaining the theory's empirical credentials in a manner sufficient to meet the neo-Kantian challenge. The next year, 1920, proved to be a pivotal one.

In the late spring of 1920, Cassirer sent Einstein the manuscript of his masterful book, *Die Einsteinschen Relativitätstheorie* (Cassirer 1921). Cassirer represented the culmination of the Marburg neo-Kantian tradition, which styled itself "critical idealism." The Marburg school started with Friedrich Albert Lange, whose *Geschichte des Materialismus* (Lange 1866) was one of the most widely read books of the era. Cassirer was trained at Marburg by Lange's successor, Hermann Cohen, with whom Einstein's philosophy teacher, August Stadler, had also trained, and Paul Natorp (see Gawronsky 1949). He shared with his teachers the hallmark Marburg thesis that the way to reconcile Kant with the developments of modern science and mathematics was to abandon the Kantian doctrine of intuition, seeking a way in which purely conceptual determinations could do the work of intuition. Kant had argued that there are only two kinds of representations, intuitions and concepts, and that they differed by virtue of the fact that intuitions are particular, whereas concepts are inherently general. The key Marburg thesis was that the accumulation of sufficiently many purely conceptual determinations can establish cognitive contact with the world in its particularity, at least in the form of a structurally unique representation. The main goal of Cassirer's 1910 book, *Substanzbegriff und Funktionsbegriff* (Cassirer 1910), was to make precisely this point informed by the best science and mathematics of the day.

In his 1921 book, Cassirer brought this framework to bear on the general theory of relativity. His main thesis, beyond the general, Marburg point about the frailty of Kant's doctrine of intuition, was that relativity theory teaches us that Kant had gone too far in asserting that Euclidean metrical structure was a priori. He argued that what is univocally fixed in the structure of general relativistic space-time is not metrical structure but topological structure, which is to say, in another idiom, the structure of intersections of world lines, or point coincidences. One might have thought that Einstein would be sympathetic. But recall Einstein's one critical response to Schlick's 1917 monograph, in which he argued that the point-event structure, hence the topological structure, of general relativistic space-time, is underdetermined by experience. No surprise, then, that Einstein, though respectful of Cassirer's efforts, still dissented from the main thesis:

> I can understand your idealistic way of thinking about space and time, and I even believe that one can thus achieve a consistent point of view. . . . I acknowledge that one must approach the experiences

> with some sort of conceptual functions, in order for science to be possible; but I do not believe that we are placed under any constraint in the choice of these functions *by virtue of the nature of our intellect.* Conceptual systems appear empty to me, if the manner in which they are to be referred to experience is not established. This appears most essential to me, even if, to our advantage, we often isolate in thought the purely conceptual relations, in order to permit the *logically* secure connections to emerge more purely. (Einstein to Cassirer, June 5, 1920, *CPAE- 10,* Doc. 44, 293)

"Conceptual systems appear empty to me, if the manner in which they are to be referred to experience is not established." There is the central problem on Einstein's view. We must first clarify the empirical interpretation of a scientific theory, such as general relativity, before we can determine whether there might be necessary structural features of the theory. No such necessity can be a consequence of the nature of cognition alone.

Einstein expanded upon this critique in a major lecture of 1921, *Geometrie und Erfarhung* (Einstein 1921). Axiomatic geometry, alone, argues Einstein, says nothing about the physical world. Only a geometry supplemented by physical definitions of geometrical primitives, like "point" and "line," what Einstein terms "practical geometry," has physical content, and the truth or falsity of Euclidean geometry, understood in this manner, is determined by experience alone. We might define the geometer's notion of "line" in terms of a physical object such as the standard meter bar in Paris, an example of what Einstein calls a "practically-rigid rod," the thermal and gravitational deformations of which are negligible for the purpose of checking whether Euclidean metrical relations obtain in physical space. In the case of general relativity, the mathematical primitive notion of the infinitesimal metric interval would, by analogy, be given a physical interpretation by means of both ideal (practically rigid) rods and ideal (practically regular) clocks.

Of course, strictly speaking, there are no perfect rods and clocks, so, as Einstein says, a "more general view" of the matter is suggested.

> Geometry (G) asserts nothing about the behavior of real things, but only geometry together with the totality (P) of physical laws. Symbolically, we may say that only the sum of (G) + (P) is subject to control by experience. Thus (G) may be chosen arbitrarily, and equally so parts of (P); all these laws are conventions. All that is necessary to avoid

> contradictions is to choose the remainder of (P) so that (G) and the totality of (P) are together in accord with experience. On this view, axiomatic geometry and the portion of the laws of nature that have been elevated to conventions appear as epistemologically equivalent. (Einstein 1921, 8)

Einstein made a similar point, only now with specific reference to neo-Kantian efforts, such as those of Cassirer, to salvage something of Kant's doctrine of a priori forms of intuition, in a 1924 review of Alfred Elsbach's *Kant und Einstein. Untersuchungen über das Verhältnis der modernen Erkenntnistheorie zur Relativitätstheorie* (Elsbach 1924):

> This does not, at first, preclude one's holding at least to the Kantian problematic, as, e.g., Cassirer has done. I am even of the opinion that this standpoint can be rigorously refuted by no development of natural science. For one will always be able to say that critical philosophers have until now erred in the establishment of the a priori elements, and one will always be able to establish a system of a priori elements that does not contradict a given physical system. Let me briefly indicate why I do not find this standpoint natural. A physical theory consists of the parts (elements) A, B, C, D, that together constitute a logical whole which correctly connects the pertinent experiments (sense experiences). Then it tends to be the case that the aggregate of fewer than all four elements, e.g., A, B, D, without C, no longer says anything about these experiences, and just as well A, B, C without D. One is then free to regard the aggregate of three of these elements, e.g., A, B, C as a priori, and only D as empirically conditioned. But what remains unsatisfactory in this is always the arbitrariness in the choice of those elements that one designates as a priori, entirely apart from the fact that the theory could one day be replaced by another that replaces certain of these elements (or all four) by others. (Einstein 1924, 1688–89)

Cassirer sought to save the Kantian project by agreeing that Kant was wrong to declare Euclidean metrical structure a priori and arguing that general relativity has taught us, instead, that a certain topological structure of space-time is a priori. This is precisely the strategy that Einstein here calls to account by faulting not the possibility of the maneuver but its arbitrariness. For Einstein, these choices are all merely conventional.

In 1920 and 1921, as Cassirer was working out his ideas on general relativity, Carnap was writing his doctoral dissertation, *Der Raum* (Carnap 1921) in Jena. Likewise inspired by general relativity's challenge to Kant, and acknowledging the influence of Natorp and Cassirer on his thinking, but also the influence of the phenomenologist, Edmund Husserl, Carnap made a threefold distinction between (a) formal, mathematical space; (b) intuitive space; and (c) physical space. Knowledge of the first is, of course, purely logical and axiomatic. Knowledge of the second, intuitive space, is synthetic a priori. Knowledge of physical space is empirical. About intuitive space, which constitutes the necessary condition for the possibility of objects of experience, Carnap argues that only a set of topological properties, not full metrical structure, is a priori.

Carnap's discussion of physical space is subtle and bears comparison with Einstein's position in *Geometrie und Erfahrung*. In brief, Carnap holds that one can freely stipulate by convention what is to count as a standard measuring rod, M ("Maßstab" = "measuring stick") and that once that is done the metrical structure of physical space R ("Raumform" = "spatial form") is determined univocally by the results of our measurements. On the other hand, Carnap says, one could proceed in the opposite manner, first freely stipulating by convention a metric structure, R, which stipulation would fix univocally the kind of measuring rod, M, necessary to yield measurement results conformable to the stipulated metrical structure. But Carnap recommends what he terms a "mediating" approach, mutually adjusting both M and R to yield the simplest overall structure for physical space. Carnap's R+M strategy has much in common with Einstein's (G)+(P) approach in that both foreground the role of conventions in theories of physical space and both hold that, given the relevant evidence, one constructs the simplest total theory consistent with that evidence.

In the context of this rich debate about the epistemological implications of general relativity and its challenge to Kant's theory of space and time, the most consequential intervention from the side of the philosophers was the publication in the fall of 1920 of Reichenbach's *Relativitätstheorie und Erkenntnis Apriori* (Reichenbach 1920). The response that this book elicited led directly to the emergence of what, today, we recognize as the classic, logical empiricist philosophy of science of the Vienna Circle and its allies in Berlin, Prague, and elsewhere. At the time of its writing, Reichenbach was still working in Berlin and auditing Einstein's lectures on general relativity, though in the spring of 1920, he took up a position as a Privatdozent at the Technische Hochschule in

Stuttgart, where he worked until summoned back to Berlin by Einstein and Planck in 1926.

That Reichenbach was in regular contact with Einstein during the book's gestation is an important part of the context. He wrote to Einstein on June 15, 1920 to ask Einstein's permission to dedicate the book to him and he added:

> You know that it is my intention with this book to point out the philosophical consequences of the theory and to show what great discoveries your physical theory has brought for epistemology. Were I to place your name at the front, I would thereby want to express what great thanks philosophy owes to you. I know quite well that only the fewest academic philosophers have a sense that, with your theory a philosophical achievement has been made and that more philosophy is contained in your physical conceptions than in all of the multivolume works of the epigones of the great Kant.[7] (Reichenbach to Einstein, June 15, 1920, *CPAE-10,* Doc. 58, 314)

Einstein graciously accepted the compliment, agreeing to the dedication. He had already expressed his enthusiasm for the book when he announced its existence in a letter to Schlick two months earlier: "The young Reichenbach has written an interesting essay on Kant and general relativity" (Einstein to Schlick, April 19, 1920, *CPAE-9,* Doc. 378, 510).

Reichenbach starts with what he takes to be an incontestable fact: General relativity has taught us that the old, Kantian doctrine of the a priori forms of intuition will not work. The empirical evidence makes it clear that we live not live in a three-dimensional, Euclidean space, but in a four-dimensional space-time with variable curvature. And yet, on Reichenbach's view, some of Kant's key insights survive the challenge from general relativity. More specifically, Reichenbach argues that Einstein's physics has taught us that we must disentangle two aspects of the a priori as it was conceived by Kant: (1) the constitutive role of the a priori in affording the grounds for the possibility of there being objects of experience, and (2) the apodictic or necessary character of the a priori. In the wake of general relativity, the latter, the claimed, necessary truth of the a priori, clearly must be abandoned. But the constitutive work of the a priori is not impugned if only we realize that the manner in which objects of scientific cognition are to be constituted changes as science advances.

In what way does the constitutive work of the a priori find expression in modern physical science? It is by means of what Reichenbach terms "principles of coordination," principles that effect a connection between theoretical terms and objects of possible experience. But whereas Kant held that the constitution of possible objects of experience was fixed univocally by virtue of the nature of the intellect, Reichenbach argues that experience teaches us which principles of coordination are best suited to the science of the day:

> If we disclaim the Kantian analysis of reason, it cannot be contested that experience contains elements that are conformable to reason. Indeed, it is precisely the principles of coordination that are determined by the nature of reason; experience merely effects the choice among all conceivable principles. All that is contested is that the rational components of knowledge are maintained independently of experience. The principles of coordination represent the rational components of empirical science. Therein lies their fundamental significance and therein are they distinguished from every, individual law, even the most general. For the individual law represents only an application of those conceptual methods that are grounded by the principle of coordination; only by means of the methods fixed by such principles do we define how the knowledge of an object is effectuated conceptually. (Reichenbach 1920, 83–84)

Today one speaks of such a view, without intending irony, as a version of the "contingent a priori."[8]

The appearance of Reichenbach's book engendered an immediate and intense debate about both the specific nature of the "principles of coordination" and whether this was such an attenuated view of the a priori as no longer to merit the label "Kantian." Schlick took up both of these points in a letter to Einstein dated October 9, 1921:

> In the last few days I have read with the greatest pleasure the booklet by Reichenbach on relativity theory and a priori knowledge. The work really appears to me to be a quite splendid contribution to the axiomatics of the theory and of physical knowledge in general. . . . Of course, in a few points I still cannot entirely support Reichenbach. . . . Reichenbach seems to me not to be fair with regard to the theory of conventions of Poincaré; what he calls a priori principles of coordination, and rightly distinguishes from the empirical principles of connection, seem to me to be wholly identical with Poincaré's "conventions" and to have no significance beyond that. R.'s reliance upon

> Kant seems to me to be, carefully considered, only purely terminological. (Schlick to Einstein, October 9, 1921, *CPAE-10*, Doc. 171, 454–55)

A couple of months later, Schlick made essentially the same points, albeit in far greater detail, to Reichenbach, himself:

> For me the presupposition of object-constituting principles is so self-evident that I have not pointed it out emphatically enough, above all in the Allg. Erkenntnisl. . . . It is quite clear to me that a perception can become an "observation" or even a "measurement" only through certain principles being presupposed by means of which the observed or measured object is then constructed. In this sense the principles are to be called a priori. . . . But there are indeed, moreover, two possibilities, that those principles are hypotheses or that they are conventions. In my opinion, precisely this turns out to be the case, and it is the central point of my letter, that I cannot discern wherein your a priori propositions are actually distinguished from conventions. That you passed over Poincaré's theory of conventions with so few words is what most amazed me about your essay. . . . The crucial places where you describe the character of your a priori principles of coordination appear to me, frankly, as quite successful definitions of the concept of convention. . . . I do not fear that you can object that conventionalism must also make use of the hypothesis that you find implicit in Kant's philosophy (p. 57) [there are no contradictory systems of principles, the hypothesis of the arbitrariness of coordinations]. Indeed, only such conventions are permitted that fit into a certain system of principles, and this system *as a whole* will be determined by experience; the arbitrariness only enters in the manner of its construction and is steered by the principle of simplicity, economy, or, as I would rather have said, the principle of the minimum of concepts. Here there appears to me to be a small gap in your essay, which is not without consequences: In the concept of knowledge you consider explicitly only the *one* side, the coordination, and you slight a little bit the other side, that the coordination should be accomplished with the fewest and consequently the most general possible concepts. (Schlick to Reichenbach, November 26, 1920, HR 015-63-22; as quoted in Howard 1994, 61–62)[9]

Reichenbach was quick to respond:

> You ask me why I do not call my a priori principles *conventions*. I believe that we will easily come to agreement about this question. Even though several systems of principles are possible, nevertheless, only one *group* of principle-systems is always possible; and precisely in this restriction there lies some knowledge. Every possible system signifies in its possibility a *property* of reality. I miss in Poincaré an emphasis on the fact that the arbitrariness of the principles is restricted, in the way one *combines* principles. For that reason I cannot adopt the name "convention." Also, we are never certain that two principles that we today allow to exist alongside one another as constitutive principles, and which are therefore both *conventions*, according to Poincaré, might not tomorrow have to be separated because of new experiences, so that between the two conventions the alternative appears as synthetic. (Reichenbach to Schlick, November 29, 1920, HR 015-63-21; as quoted in Howard 1994, 62)

Schlick replied that Poincaré did understand this point:

> (1) on the question of the "conventions." If Poincaré did not explicitly emphasize that conventions are not independent of one another, but are always possible only as groups, still one would naturally do him quite an injustice, if one believed, that he was not aware of this circumstance. This was obviously the case, and he would have repudiated with mockery the nonsense that, e.g., Dingler has perpetrated with the concept of conventions while misunderstanding this circumstance. Thus, in my view, nothing stands in the way of the retention of the term. (Schlick to Reichenbach, December 11, 1920, HR 015-63-19, as quoted in Howard 1994, 62)

The key point for Schlick in all of this was that contingent a priori principles of coordination were better viewed as mere conventions. He stressed this in a note about Reichenbach's book that he appended to his review of Cassirer's *Zur Einsteinschen Relativitätstheorie* (Schlick 1921) and in his more extensive review of Reichenbach's book in *Die Naturwissenschaften* the following year (Schlick 1922). As we will see, Reichenbach soon conceded the point, and their together achieving this clarification and consensus became the turning point in the development of the new kind of empiricism that today we know well as the mature, logical empiricist, verificationist form of empiricism that dominated the philosophy of science literature for much of the rest of the twentieth century. In his review of Cassirer's book, Schlick had heralded the birth of such a new empiricism, one not identical with the "sensualism" of Mach. Schlick argued that only by its lights could the empirical integrity of general relativity be maintained against the challenge presented by neo-Kantian responses that, as with Cassirer's, sought to salvage a role for at least an attenuated a priori in the face of general relativity's seemingly empirically driven dissent from Kantian orthodoxy about the necessary, a priori role of space and time in scientific cognition (Schlick 1921, 323). What was needed of this new form of empiricism was a clear explanation of how, in spite of there being a conventional element in all empirical science, the crucial implication about the metrical structure of general relativistic space-time was objectively grounded in empirical fact. Getting clear about the precise role of conventions in science, namely, in the form of coordinating principles or, as they later will be termed, coordinating definitions was the crux of the issue.

Schlick had, of course, been pondering the role of conventions in science ever since his first essay on the philosophical significance of the theory of relativity and his 1917 monograph. But the question occupied center stage in Schlick's *Allgemeine Erkenntnislehre,* in which he sought to develop the germ of a scientific epistemology in the 1915 and 1917 works into a general philosophy of science (Schlick 1918). While not directly derived from Duhem, Schlick's understanding of conventions in science during these years had much in common with Duhem's holist, underdeterminationist version of conventionalism, meaning that, in principle, any element of scientific theory could be regarded as conventional or, rather, that every proposition in a theory partook of the conventional nature of the entire theory, it being only whole theories that are tested empirically, not individual propositions, one by one.

Duhem had argued that there was no principled, systematic way to distinguish between definitions and properly empirical claims, with the moment of convention restricted to the former. That was the view of Poincaré (Poincaré 1902a, 1902b, 110–34) and Edouard Le Roy (Le Roy 1901), and Duhem registered his strong disagreement in *La Théorie physique: son objet et sa structure* (Duhem 1906), specifically disputing Le Roy's claimed example according to which the law of acceleration was insulated against empirical refutation because it had the status of a purely conventional definition of "free fall." Duhem argued, by contrast, that we could so regard the law of acceleration or we could choose to regard it as an empirically testable claim, the choice between these two being only a matter of simplicity and convenience (Duhem 1906, 208–16).

In much the same way, Schlick argued in 1918 that the distinction between definitions and what he then called "judgments" was not a logical or epistemological one, but pragmatic or "relative," and he emphasized that determining how to parse the propositions in a theory between definitions and empirical assertions is, in the end, a matter of convenience and simplicity:

> Once a science has developed into a rounded-out more or less closed structure, what is to count in its systematic exposition as definition and what as knowledge [Erkenntnis] is no longer determined by the accidental sequence of human experiences. Rather, those judgments will be taken as definitions that resolve a concept into the characteristics from which one can construct the greatest possible number (possibly all) of the concepts of the given science in the simplest possible manner. (Schlick 1918, 50)

But by the time of his exchange with Reichenbach, Schlick had come to realize that, if the distinction between proper empirical judgments and definitions were merely a relative one, then the empiricist faces a serious problem. The dogged defender of a claim seemingly threatened by general relativity, such as the assertion that the Euclidean metrical form was the necessary form of outer intuition, could exploit that very relativity to argue that such an assertion must be regarded as functioning in a definitional manner, thus insulating it from the threat of empirical refutation, just as Poincaré and Le Roy saw the laws of mechanics as being so insulated. In other words, the defender of Kant could argue that, since there is nothing inherently empirical in any judgment, then the putatively a priori status of a claim about the metrical structure of space suffices to mark off such a claim as a constructive or constitutive principle, one defining, by construction, the objects of scientific cognition. Being thus, that judgment becomes a fixed element in scientific theory, one that cannot be impugned by the results of observation and experiment.

Recall that we earlier found Einstein invoking this same Duhem-Schlick point of view, according to which the distinction between definitions and empirical judgments is merely relative as, ironically, the definitive argument against a neo-Kantian philosophy of science, holding that it was, in fact, the very arbitrariness of the distinction that undermined the Kantian position. Einstein argued that, in order to test a theory, we must choose to hold some elements of the theory fixed while regarding the other elements as variable in the light of experience. But since there was no logical basis for parsing the theory in one way rather than another, that provisional fixing of some elements as a matter of practical necessity could not be taken as reason for regarding those provisionally fixed elements of theory as a priori (Einstein 1921, 1924).

But here is Schlick, at the very same time, worrying that a merely relative distinction between definitions and judgments would not afford the empiricist a sufficiently strong response to the neo-Kantians. We will find that, by the end of the 1920s, Einstein and Schlick will have parted company, philosophically, over precisely this point.

It was not only Schlick but also Reichenbach who had come to fear that a Duhemian, holist, underdeterminationist form of conventionalism left an opening for the neo-Kantian. Here is how he voiced that concern in a 1922 paper, "Die gegenwärtige Stand der Relativitäsdiskussion":

> One should not contend that, since the logical equivalence of all geometries leaves the internal consistency of Euclidean geometry unaffected, the validity of synthetic *a priori* judgments remains unchallenged; such an appeal to conventionalism is denied to Kantianism. The internal consistency of geometry is *analytic*. Conventionalism would admit all *logically consistent* conceptual systems as possible structural forms of empirical knowledge, but the significance of synthetic *a priori* principles consists in the fact that they constitute a specific choice among the *logical* possibilities. Similarly, the law of causality is not *logically* necessary, since uncaused events are logically possible; according to Kant, it is a synthetic *a priori* principle which excludes uncaused events. If "synthetic *a priori*" meant nothing but "internally consistent," Kantians would have to admit that, at some future time, we might gain knowledge of uncaused events. But on such an interpretation, synthetic *a priori* principles would ultimately degenerate into empty formulas that would impose no limits on experience. This is the reason why eliminating the metric from pure intuition leads to a denial of synthetic judgments *a priori*. (Reichenbach 1922, 30)

What was going to be needed to block the neo-Kantian exploitation of the conventionalist maneuver? What would be needed was a way of arguing that, in spite of the role of conventions in science, specific, individual, empirical propositions, such as general relativity's assertion about the metrical structure of space-time, possessed their own, individual, empirical content and so were to be regarded as being tested not as part of a whole body of theory but one by one. In other words, one had to eliminate the holism from conventionalism. How was that to be done?

The answer turned out to be that one restricted the role of conventions to analytic coordinating definitions, the stipulation of which fixed a determinate, empirical meaning for every primitive term in a theory and thereby also fixed a determinate, empirical meaning for every synthetic, empirical proposition in the theory, rendering each one of those empirical propositions susceptible to direct empirical testing. Should a claim about the metrical structure of some local region of space-time pass the test, as with Arthur Eddington's confirmation of Einstein's claim about the bending of the path of light passing near the Sun, then that claim stands as empirically verified fact, no longer deniable by the neo-Kantian. And therein lies the birth of verificationism in logical empiricist philosophy of science.

Reichenbach was the first to give clear expression to this new, verificationist version of empiricism in his 1924 book, *Axiomatik der relativistischen Raum-Zeit-Lehre* (Reichenbach 1924), and it became still better known through his widely read and influential 1928 book, *Philosophie der Raum-Zeit-Lehre* (Reichenbach 1928). The analysis in the 1924 book begins with what is a principled and rigid distinction between "axioms," which Reichenbach terms "elementary facts" that are "based" on experiments (Reichenbach 1924, 6), and "coordinative definitions, which he introduces thus:

> Definitions are arbitrary; they are neither true nor false. They are merely to be analyzed with respect to their logical properties, their uniqueness, consistency, and, under certain conditions, their simplicity. It is characteristic of the axiomatization of physics compared to that of mathematics that there exists such a distinction between axioms and definitions; an essential task of the axiomatization consists in tracing this distinction within the theoretical system.
>
> However, even definitions in physics are different from definitions in mathematics. The mathematical definition is a *conceptual definition*, that is, it clarifies the meaning of a concept by means of other concepts. The physical definition takes the meaning of the concept for granted and coordinates to it a physical thing; it is a *coordinative definition*. Physical definitions, therefore, consist in the coordination of a mathematical definition to a "piece of reality"; one might call them *real definitions*. (Reichenbach 1924, 7–8)

It is these coordinative definitions that fix the determinate, empirical content of the concepts employed in the formulation of the axioms and, as a consequence, the empirical contents of the axioms or "elementary facts."

How did Schlick announce his own conversion to this new, verificationist version of empiricism? His first, extended presentation of his new view is found in the 1925 second edition of his *Allgemeine Erkenntnislehre* (Schlick 1925), to which he added several new sections that mainly deal precisely with the role of conventions, definitions, and empirical judgments in scientific theory. Nowhere in the second edition does he flatly assert that he has changed his mind specifically about whether the distinction between definitions and empirical judgments is a relative one, as he had argued in the 1918 first edition, leaving the reader with the somewhat misleading impression that the new material in the second edition seeks merely to clarify Schlick's older views. But the shift in Schlick's views is nonetheless dramatic, as a comparison of the two editions makes clear.[10]

Consider the new section 11 of the second edition, "Definitionen, Konventionen, Erfahrungsurteile." Schlick here asserts explicitly that scientific theories contain only conventional, analytic definitions and synthetic empirical judgments, thus no synthetic a priori judgments, and that the distinction between definitions and empirical judgments is not relative:

> We might be tempted to think that the distinction between analytic and synthetic judgments cannot be drawn sharply, since one and the same judgment may be synthetic or analytic depending on what we include in the subject concept. But this opinion ignores the fact that the judgment is really *not* the same in the two cases. In the first case, we define the concept *body* in "All bodies are heavy" so that being heavy is one of its features; in the second case, we do not. True, the sentence contains the same *words* each time, but they designate different judgments, for the word "body" has a different meaning in each. We explained above (§ 8) that one and the same (linguistic) sentence can express both a definition and a piece of knowledge. It all depends on what concepts we connect to the words. The partitioning of judgments into analytic and synthetic is thus something quite well defined and objectively valid, and does not depend, say, on the subjective standpoint or mode of comprehension of the one who judges. (Schlick 1925, 76)

How did Schlick try to cover the fact that his views had changed in important ways out of the felt need to block the Kantian exploitation of the conventionalist strategem? He now argues that it all has to do with the difference between "ideal"—meaning "formal," and "real"— meaning "empirical," sciences:

> Every judgment we make is either definitional or cognitive. This distinction, as we noted above (§ 8), has only a relative significance in the conceptual or "ideal" sciences. It emerges all the more sharply, however, in the empirical or "real" sciences. In these sciences it has a fundamental importance; and a prime task of epistemology is to make use of this distinction in order to clarify the kinds of validity possessed by various judgments. (Schlick 1925, 69)

What is the point of now making the distinction between definitions and judgments in "real" theories a fixed, not relative one, with the moment of convention in scientific theory now restricted to definitions? It is precisely to guarantee that experience alone determines the truth or falsity of synthetic, empirical judgments: "Once a certain number of concepts are fixed by convention, the relations that hold between the objects so designated are not conventional. They must be determined through experience" (Schlick 1925, 72). Or, as Schlick writes a few pages later:

> The system of definitions and cognitive judgments, which constitutes any real science, is brought into congruence at individual points with the system of reality, and is so constructed that congruence then follows automatically at all remaining points. . . . If the whole edifice is correctly built, then a set of real facts corresponds not only to each of the starting points—the fundamental judgments—but also to each member of the system generated deductively. Every individual judgment in the entire structure is uniquely coordinated to a set of real facts. (Schlick 1925, 78)

And Schlick clearly identifies Kantian a priorism as the alternative to be blocked by this new form of empiricism:

> According to him [Kant], besides the two classes of judgments we have described—definitions in the widest sense (Kant calls them analytic

judgments) and empirical judgments or hypotheses (these he calls synthetic judgments *a posteriori*)—there is a third class, the so-called synthetic judgments *a priori*. . . . The fact of the matter is that no one has as yet succeeded in exhibiting a synthetic judgment *a priori* in any science. That Kant and his followers nevertheless believed in their existence may be explained quite naturally by the fact that among both the definitions and the empirical propositions of the exact sciences we find statements that are deceptively similar to synthetic judgments *a priori*. In the class of definitions, which by their very nature possess a validity independent of experience and thus are *a priori*, there are a great many conventions that, viewed superficially, seem not to be derivable from definitions and hence to be synthetic. Their true character as conventions is revealed only by a most painstaking analysis. . . .

Once we demonstrate . . . that the judgments held to be synthetic and *a priori* are in fact not synthetic or not *a priori*, there is no reason whatever to suppose that judgments of this strange sort might yet exist in some obscure corner of the sciences. And this is sufficient ground for us to try in what follows to explain all knowledge of reality as a system built up exclusively of judgments belonging to the two classes described above. (Schlick 1925, 73–75)

What kinds of conventions might at first glance seem not to be conventional because, being not obviously derived from definitions, could be mistaken as synthetic judgments? In the case of general relativity it would be the coordinating definitions that associate fundamental mathematical-physics concepts, such as the line element, with rods and clocks. Those are not lexical definitions and, so, might be wrongly regarded as synthetic. But, in fact, they are conventional, analytic, coordinating definitions, and, as per Schlick's new account, once fix via coordinating definitions one's rods and clocks and all remaining synthetic propositions acquire determinate empirical content such that their truth or falsity can be determined unambiguously on the basis of the corresponding experience. From this time forward, until his tragic assassination in 1936 by a deranged former student, Schlick consistently defended this verificationist logical empiricism that became the principal epistemological legacy of the Vienna Circle well into the latter half of the twentieth century.

The Two Narratives Converge: "*Eindeutigkeit,*" Definitions, Conventions, and Empirical Content

We have been following two historical narratives. One derives mainly from debates about the hole and point coincidence arguments that concerned primarily the objective, ontological content of general relativistic space-time and led to the view that point coincidences in such forms as the intersections of world lines constituted the basis of that ontology because of their diffeomorphism invariance and, hence, their being fixed univocally. The other derives from debates in the late 1910s and early 1920s with neo-Kantians of various kinds about whether the question of the metrical structure of general relativistic space-time is based exclusively on experience or might contain some a priori component. This line of discussion eventuated in the emergence in the early 1920s of the verificationist logical empiricism that dominated the philosophy of science literature for decades thereafter. We noted along the way a confusion in Schlick's thinking about the relationship between point coincidences, which are ontologically privileged because of their univocality, their "Eindeutigkeit," and what has been termed "pointer coincidences," such as the coincidence of a pointer on an instrument and a point on a scale, which are held by Schlick to be epistemically privileged by their objectivity, which he regarded as derivative from their also being point coincidences. But this was a spurious connection, as Reichenbach clearly explained.

The two narrative threads do converge in a more significant way, however, in the latter 1920s, in the work of Rudolf Carnap, who, since his 1921 doctoral dissertation, *Der Raum,* had been thinking carefully about the relationship between the univocal determination of physical or ontological structure in space-time theory and the univocal determination of epistemic structure in what, in the dissertation, he termed "intuitive space." The convergence appears in an interesting and important but, lamentably, little known essay of Carnap's from 1927, "Eigentliche und uneigentliche Begriffe," which was written at the same time as Carnap was working on the final version of his well-known book, *Der logische Aufbau der Welt* (Carnap 1928).

While Carnap would later part company with Schlick and Reichenbach over their defense of verificationist logical empiricism, in 1927 he was still in broad agreement with them. Carnap defines "eigentliche Begriffe," or "proper concepts," as explicitly defined and, thus, categorical or univocal concepts. "Uneigentliche Begriffe," or "improper concepts," by contrast, are implicitly defined, in Hilbert's sense, by their systematic role in the formalized theory in question and are therefore noncategorical, lacking univocality. Carnap allows that noncategorical, implicitly defined, improper concepts commonly play a role in the deeper parts of a theory, as, perhaps, in the fundamental axioms. But he argues that, in order for a theory to possess determinate, univocal empirical content and, thus, to be testable in the manner required by logical empiricist philosophy of science, all of the empirical primitive terms in the theory must be explicitly defined, univocal, proper concepts. Coordinating definitions afford precisely that. Carnap explains:

> In the systematic construction of the knowledge of reality, real concepts are constituted step-by-step. As a term in this construction, every real concept possesses an immediate relation to reality. By contrast, improper concepts so-to-say hover in mid-air. They are introduced through an *AS* [axiom system] that does not refer directly to reality. The axioms of this *AS* and the theorems deduced from them do not properly constitute a theory (since, indeed, they deal with nothing determinate), but only a theory-schema, an open form for possible theories. But if, in the system of knowledge, a real concept appears for which it can be shown empirically that it has the formal makeup indicated in the *AS* for the improper concept, then the *AS* has found a realization: In place of the improper concept, which is just a variable, the real concept can be substituted. Thus, the structures of physical space (points, lines, etc.) evince empirically the makeup that the axioms of geometry express for "points" (in the improper sense), etc. . . . Through this contact between the real concept and the axioms (the former satisfy the latter), the connection to the entire theory-schema based on the *AS* is accomplished in one stroke. The blood of empirical reality streams in through this one point of contact and flows into the most highly branched veins of the heretofore empty schema, which is thereby transformed into a filled-out theory. (Carnap 1927, 372–73)

As mentioned above, the notion that Carnap uses here, "categoricity," was first introduced by Hilbert in his work on the axiomatization of Euclidean geometry and had, by 1927, been the focus of a long discussion involving logicians, mathematicians, and philosophers as different notions of univocality were discerned. Carnap himself was to make important contributions in this arena, especially as regards distinguishing the notion of semantic completeness or categoricity from the concepts of deductive completeness and decidability in formal systems (see, e.g., Carnap 1929). While these developments in logic, set theory, and the foundations of mathematics, which led to the establishment of the Gödel incompleteness theorems, are ancillary to the issues of primary concern in tracking the influence of general relativity on the development of twentieth-century philosophy of science, it is still noteworthy that, however much the point might not be appreciated in the literature on concepts of completeness in the study of formal systems, the roots of that history also go back, in part, to concerns that arose in the effort to make clear, philosophical sense of general relativity and to defend its empirical integrity in the face of Kantian challenges.[11] The influence of general relativity on twentieth-century thought extends far beyond just physics and the philosophy of science.

Conclusion: A Parting of the Ways

Einstein was, of course, an important contributor to the history that has been surveyed in this chapter. He was not merely a helpful interlocutor for Schlick, Reichenbach, Frank, Cassirer, and other, major contributors to the development of twentieth-century philosophy of science, such as Weyl. He was also one of the era's most important philosophers of science in his own right (see Howard 2005). Philosophically and personally, he was very close to both Schlick and Reichenbach during the crucial years in which the classic, verificationist version of logical empiricism was being developed in response to Kantian critiques of general relativity. It might seem surprising, therefore, that, at the very time when modern logical empiricism was being refined and articulated, Einstein began to lose his enthusiasm for the very philosophy of science whose birth he facilitated.

A clear sign of the gap that had opened is to be found in a letter from Einstein to Schlick in late 1930. Commenting on the manuscript of a paper by Schlick on causality (Schlick 1931), Einstein writes:

> From a general point of view, your presentation does not correspond to my way of viewing things, inasmuch as I find your whole conception, so to speak, too positivistic. Indeed, physics *supplies* relations between sense experiences, but only indirectly. For me *its essence* is by no means exhaustively characterized by this assertion. I put it to you bluntly: Physics is an attempt to construct conceptually a model of the

> *real world* as well as of its law-governed structure. To be sure, it must represent exactly the empirical relations between those sense experiences accessible to us; but *only* thus is it chained to the latter. . . . You will be surprised at the "metaphysician" Einstein. But every four- and two-legged animal is de facto in this sense a metaphysician. (Einstein to Schlick, November 28, 1930, EA 21-603)

The issue here is not just that Einstein, the realist, had grown disenchanted with logical empiricism after it had taken an anti-metaphysical turn (see Howard 1993). Indeed, the problem is that Einstein had come to dissent strongly from the very "new empiricism" that Schlick and Reichenbach had developed in their efforts to shield general relativity from Kantian critiques. The point of difference was verificationism itself.

Since at least 1909, when Einstein appears first to have read the German translation of Duhem's *La Théorie physique: son objet et sa structure* in the German translation by his Zurich friend and neighbor, Friedrich Adler (Duhem 1908), he had evinced enthusiasm for Duhem's holist and underdeterminationist version of conventionalism, alluding to it in his lectures on electricity and magnetism at the University of Zurich in the Winter semester of 1910–1911 (see Howard 1990, 1993). We saw above how Einstein deployed Duhemian holism in his own, rather different reply to the neo-Kantians in his 1924 review of Alfred Elsbach's book, *Kant und Einstein*, where, echoing Duhem's critique of Poincaré's version of conventionalism, Einstein had argued that there was no principled basis for distinguishing fixed, a priori elements of theory from contingent, empirical claims (Elsbach 1924; Einstein 1924). But the Schlick-Reichenbach version of verificationist logical empiricism was squarely based, as we have seen, on the premise that there was a fundamental and principled difference in kind between analytic coordinating definitions and synthetic, empirical claims, the moment of convention in science being restricted to the former. It was that claim, in particular, from which Einstein dissented. Here is how Einstein put the point in his important 1936 essay, "Physik und Realität":

> We shall call "primary concepts" such concepts as are directly and intuitively connected with typical complexes of sense experiences. All other concepts are—from the physical point of view—meaningful only insofar as they are brought into connection with the "primary concepts" through statements. These statements are partly definitions of the concepts (and of the statements logically derivable from them)

> and partly statements that are not derivable from the definitions, and that express at least indirect relations between the "primary concepts" and thereby between sense experiences. Statements of the latter kind are "statements about reality" or "laws of nature," i.e., statements that have to prove themselves on the sense experiences that are comprehended in the primary concepts. Which of the statements are to be regarded as definitions and which as laws of nature depends largely upon the chosen representation; in general it is only necessary to carry through such a distinction when one wants to investigate to what extent the whole conceptual system under consideration really possesses content from a physical standpoint. (Einstein 1936, 316)

It is ironic that the position that Einstein here defends calls to mind nothing so much as Schlick's own, later repudiated assertion in the 1918 first edition of his *Allgemeine Erkenntnislehre* that the distinction between definitions and empirical judgments was merely "relative."

Why did Einstein care so much about this issue? One possibility is that the later Schlick-Reichenbach position, according to which each, individual, empirical proposition possesses a determinate empirical content such that the corresponding experience determines univocally the truth or falsity of that proposition, would have been incompatible with Einstein's often asserted and deeply held view that theories and the fundamental concepts that they comprise were "free creations of the human mind" (see, e.g., Einstein 1921, 234). But that the issue was extremely important to Einstein is clear from his exchange with Reichenbach in the pages of the 1949 Library of Living Philosophers volume, *Albert Einstein: Philosopher-Scientist*. Reichenbach had contributed to this collection an essay on "The Philosophical Significance of the Theory of Relativity" (Reichenbach 1949). In it, he recapitulated the verificationist version of logical empiricism that he and Schlick had developed in the 1920s and that he was to defend until his death in 1953:

> Another confusion must be ascribed to the theory of conventionalism, which goes back to Poincaré. According to this theory, geometry is a matter of convention, and no empirical meaning can be assigned to a statement about the geometry of physical space. Now it is true that physical space can be described by both a Euclidean and a non-Euclidean geometry; but it is an erroneous interpretation of this relativity of geometry to call a statement about the geometrical structure of physical space meaningless. The choice of a geometry is arbitrary

> only so long as no definition of congruence is specified. Once this definition is set up, it becomes an empirical question *which* geometry holds for a physical space. . . . The combination of a statement about a geometry with a statement of the co-ordinative definition of congruence employed is subject to empirical test and thus expresses a property of the physical world. The conventionalist overlooks the fact that only the incomplete statement of a geometry, in which a reference to the definition of congruence is omitted, is arbitrary; if the statement is made complete by the addition of a reference to the definition of congruence, it becomes empirically verifiable and thus has physical content. (Reichenbach 1949, 297)

Einstein's reply is both amusing and enlightening. It begins as a dialogue between "Reichenbach" and "Poincaré," but, owing to his "respect" for Poincaré, Einstein replaces the latter with an "anonymous non-positivist," who first gets "Reichenbach" to agree that, as Einstein had long before argued (Einstein 1921), geometry is tested not alone but only together with physics, and then says this:

> *Non-Positivist*: If, under the stated circumstances, you hold distance to be a legitimate concept, how then is it with your basic principle (meaning = verifiability)? Must you not come to the point where you deny the meaning of geometrical statements and concede meaning only to the completely developed theory of relativity (which still does not exist at all as a finished product)? Must you not grant that no "meaning" whatsoever, in your sense, belongs to the individual concepts and statements of a physical theory, such meaning belonging instead to the whole system insofar as it makes "intelligible" what is given in experience? Why do the individual concepts that occur in a theory require any separate justification after all, if they are indispensable only within the framework of the logical structure of the theory, and if it is the theory as a whole that stands the test? (Einstein 1949, 678).

This passage is remarkable for many reasons, not least of which is that it anticipates by two years Quine's far better known extension of Duhemian theory holism to the semantic holism that became a famous centerpiece of Quine's philosophy of language (see Quine 1951, 1960). In "Two Dogmas of Empiricism, "Quine named as "reductionism" the very same verificationism that Einstein here so delightfully repudiates.

Quine's critique of verificationism is one of the signal moments in the history of twentieth-century philosophy of science. Can there be any better proof of the sophistication and significance of Einstein's contributions to that history and of the importance of general relativity for understanding the fine structure of twentieth-century philosophy of science than the fact that Einstein scooped Quine by two years?[12]

Notes

Kormos Buchwald: Preface

1. Einstein to G. E. Hale, October 14, 1913, published in the *Collected Papers of Albert Einstein* (*CPAE*), Vol. 5, Doc. 477.
2. G. E. Hale to Einstein, November 8, 1913, *CPAE* Vol. 5, Doc. 483.
3. See especially the introductions to *CPAE* Vols. 14 and 15.

Barish: The Quest for (and Discovery of) Gravitational Waves

1. Abbott et al. (2016a).
2. Abbott et al. (2016b).

Thorne: One Hundred Years of Relativity: From the Big Bang to Black Holes and Gravitational Waves

1. Several such movies are in the online supplementary data of James, von Tunzelmann, Franklin, and Thorne (2015).
2. For far greater detail on my personal historical perspective, see Thorne (2019).

Buonanno: The New Era of Gravitational-Wave Physics and Astrophysics

* This chapter reviews scientific results of the LIGO Scientific Collaboration updated through Fall 2016. It includes some material that the author originally published in *CERN Courier* (January /February 2017, pp. 19–20).

1. Abbott et al. (2016a).
2. Abbott et al. (2016b).
3. Abbott et al. (2016c).
4. See, for example, Buonanno and Sathyaprakash (2015), and Choptuik, Lehner, and Pretorius (2015), and references cited therein.
5. Abbott et al. (2016d).
6. Abbott et al. (2016e).
7. Abbott et al. (2016c); Abbott et al. (2016f).
8. Abbott et al. (2016c).
9. Since Fall 2016, Advanced LIGO and Virgo announced four other gravitational-wave detections, notably three binary black holes (GW170104, GW1170608, GW170814) and one binary neutron star (GW170817). The latter heralded the era of multimessenger astronomy.
10. This paradigm was solved with the observation of the first binary neutron-star merger, GW170817.
11. Armano et al. (2016).

Kennefick: The Wagers of Science

* I would like to thank Jed and Diana Buchwald for their invitation to present this chapter and Dennis Lehmkuhl for intensive and enjoyable discussions on the material relating to numerical relativity, especially in its historical and philosophical aspects and for direct contributions to the text in that section.

1. For materials related to the original announcement of GW150914, including the waveform discussed below, see https://www.ligo.caltech.edu/news/ligo20160211.
2. This version of the figure can be viewed on the LIGO website at https://www.ligo.caltech.edu/image/ligo20160211a.
3. In fairness, I have since spoken to some of the younger people present that day, such as Bernd Bruegmann, who insist that they were quite optimistic in accepting Kip's wager.
4. For a thorough discussion of this controversial history, see Kennefick (2007).
5. The letter is dated February 19. The paper announcing the discovery of gravitational waves was published on February 11, 2016, just in time for the centenary of Einstein's first written reference to the subject.
6. The last line of the paper's abstract reads, "Theorists must begin now to lay a foundation for extracting the waves' Information" (referring to extracting information from the signals in the LIGO instruments).
7. I am occasionally embarrassed when it is pointed out in public (for instance when I am being introduced before a lecture) that I am a co-author of this famous paper. The truth is that I contributed nothing to it. Early drafts of the paper contained only two footnotes that referenced work to which I contributed. As the long process of getting the paper by referees at the journal proceeded, first one and then the other of these footnotes was dropped. At this point there was no longer any justification for keeping my name on the author list, but I was too shy to point this out to anyone. Since the matter was too insignificant to be noticed by anyone else, my name remained in the author list by sheer inertia. It is by far the most cited article with my name on it.
8. Smarr and others submitted an unsolicited proposal to the NSF, known as the black proposal, from its cover, which inspired the founding of the NCSA and other centers. It was at the NCSA that the first graphical web browser, Mosaic, was written in the 1990s. The black proposal can be read online at http://www.ncsa.illinois.edu/20years/timeline/documents/blackproposal.pdf.
9. The similarity to many web browsers, many of them descended from NCSA Mosaic, which operate on a similar philosophy of using "plug-ins," is obvious.

Renn: The Genesis and Transformation of General Relativity

* I am deeply grateful to Diana Kormos Buchwald and Jed Buchwald for their generous hospitality and friendship during my stay at Caltech in early 2016, as well as for numerous stimulating discussions and their sustained encouragement of my research. I am also grateful to the team of *The Collected Papers of Albert Einstein* at CalTech for many helpful hints and friendly discussions. Special thanks also go to Hanoch Gutfreund and the Albert Einstein Archives of the Hebrew University Jerusalem for their unfailing support, as well as to my colleagues Alexander Blum, Jean Eisenstaedt, Hanoch Gutfreund, Michel Janssen, Roberto Lalli, and Matthias Schemmel for a critical reading of the manuscript and helpful suggestions, and, last but not least, to Lindy Divarci for her careful and thorough editing of the text, and to Laurent Taudin for his wonderful illustrations.

1. Abbott et al. (2016).
2. This chapter is an extended version of a talk given at CalTech during a research stay on the occasion of the Francis Bacon Award in the History and Philosophy of Science and Technology. The following is based on joint research with a team of colleagues, including

Alex Blum, Diana Buchwald, Olaf Engler, Hanoch Gutfreund, Michel Janssen, Roberto Lalli, Christoph Lehner, John Norton, Robert Rynasiewicz, Don Salisbury, Tilman Sauer, Matthias Schemmel, John Stachel, and Kurt Sundermeyer.

3. For more detailed accounts, see Renn (2007b); Janssen and Lehner (2014); Gutfreund and Renn (2015b, 2017); Lalli (2017); Blum et al. (2017a), as well as the literature given in the following.
4. For the following, see Renn and Rynasiewicz (2014) and Norton (2014).
5. Einstein (1907).
6. Einstein (after 1920).
7. For the following, see Renn (2007a); Janssen (2014a, 2014b).
8. Einstein (after 1920) in *CPAE* 7E, Doc. 31, 136.
9. See Renn (2007a) and Renn and Rynasiewicz (2014).
10. Gutfreund and Renn (2015a, 74).
11. Mach (1989 [1883]). For a discussion of its influence on Einstein, see Renn (2007d). See also Hoefer (1994).
12. Einstein (after 1920) in *CPAE* 7E, Doc. 31, 136.
13. This formulation goes back to John Archibald Wheeler (1998, 235).
14. See Einstein to Conrad Habicht, December 24, 1907, in *CPAE* 5E, Doc. 69, 47, and the discussion in Gutfreund and Renn (2015b), 26.
15. Interview of Otto Stern by Res Jost, 1961: tape recording, ETH-Bibliothek Zürich, Bildarchiv (Sign.: D 83:1–2). Cited in Toennies et al. (2011, 1074).
16. For commentaries, see appendix A in *CPAE* 3; Buchwald, Renn, and Schlögl (2013); Blum et al. (2012).
17. For a discussion of the role of the rotating disk, see Stachel (1989b) and Janssen (2014b). See also the discussion in Gutfreund and Renn (2017), 31–32.
18. Sauer (2014), 5.
19. Wheeler (1998), 235.
20. See Hentschel (1997).
21. Einstein to Erwin Freundlich, September 1, 1911. *CPAE* 5E, Doc. 281, 201.
22. Einstein (1936). See Renn, Sauer, and Stachel (1997); Renn and Sauer (2003); Sauer (2008).
23. The Zurich Notebook was the object of an extended research project (following a pioneering article by John Norton [1984]), which was executed jointly by Michel Janssen, John Norton, Jürgen Renn, Tilman Sauer, and John Stachel. The results have been published in the four-volume work *The Genesis of General Relativity* (Renn 2007b). For succinct accounts, see Renn (2004), Janssen and Renn (2015a, 2015b, forthcoming).
24. Einstein to Arnold Sommerfeld, 29 October 1912, *CPAE* 5E, Doc. 421, 324.
25. Einstein and Grossmann (1913).
26. Stachel (2007).
27. Janssen and Renn (2015a). See also Renn and Sauer (1999).
28. For the following, see Renn (2005) and Renn and Sauer (2007).
29. Renn (2005).
30. Renn (2007b).
31. For the following, see Renn (2005) and Janssen and Renn (2015a).
32. Einstein to Elsa Löwenthal, 23 March 1913, *CPAE* 8E, Doc. 434, 331.
33. Einstein ([1949] 1979, 85). For an extensive discussion, see Gutfreund and Renn (2020).
34. van Dongen (2010).
35. "On the Methods of Theoretical Physics," Herbert Spencer lecture, held in Oxford on June 10, 1933. Einstein (1954), 270.
36. "Editorial Note: The Einstein-Besso Manuscript on the Motion of the Perihelion of Mercury," in *CPAE* 4, 344–59; Earman and Janssen (1993); Janssen and Renn (forthcoming).
37. Einstein (1915).
38. David Hilbert to Einstein, November 19, 1915, *CPAE* 8E, Doc. 149, 149.
39. The prediction of the *Entwurf* theory (18″ per century) had been published by Johannes Droste (1914).

40. Einstein to Paul Ehrenfest, May 24, 1916, *CPAE* 8E, Doc. 220, 213. See Janssen and Renn (2015a, 2015b).
41. Einstein and Grossmann (1914). See also Einstein (1914).
42. The crucial final steps from the *Entwurf* theory to general relativity have been reconstructed in Janssen and Renn (2007) on which the following is based.
43. See Janssen and Renn (2015b). See also Renn (2006, 2007b [vols. 3 and 4], forthcoming) and Renn and Schemmel (2012).
44. Einstein to Michele Besso, after January 1, 1914, in *CPAE* 8E, Doc. 499, 374.
45. For discussions, see Stachel (1989a, 2014), Janssen (2014b), and see also the discussion in Gutfreund and Renn (2015b), 25.
46. Einstein, November 4, 1915, "On the General Theory of Relativity," in *CPAE* 6E, Doc. 21, 98–107.
47. Einstein, November 11, 1915, "On the General Theory of Relativity (Addendum)," in *CPAE* 6E, Doc. 22, 108–10.
48. Einstein, November 18, 1915, "Explanation of the Perihelion Motion of Mercury from the General Theory of Relativity," in *CPAE* 6E, Doc. 24, 112–16.
49. Einstein, November 25, 1915, "The Field Equations of Gravitation," in *CPAE* 6E, Doc. 25, 117–20.
50. Einstein, November 4, 1915, "On the General Theory of Relativity," in *CPAE* 6E, Doc. 21, 102; Einstein to Arnold Sommerfeld, November 28, 1915, *CPAE* 8E, Doc. 153, 152–53.
51. Janssen and Renn (2007).
52. This is strikingly illustrated also by large overlap between the paper Einstein wrote in 1914 on the *Entwurf* theory, "The Formal Foundation of the General Theory of Relativity" (Einstein 1914) and a paper he wrote in 1916, "Hamilton's Principle and the General Theory of Relativity" (Einstein 1916b). See Janssen and Renn (2007, 903–9). See also Sauer (2005b).
53. Einstein to Arnold Sommerfeld, November 28, 1915, *CPAE* 8E, Doc. 153, 152.
54. Einstein to Paul Ehrenfest, December 26, 1915, *CPAE* 8E, Doc. 173, 167.
55. Hilbert (1915).
56. For discussion, see Corry, Renn, and Stachel (1997), Renn and Stachel (2007), and Sauer (2005a).
57. See Mie (1913); see also Born (1914).
58. This is discussed in Renn and Sauer (2007) and Janssen (2014b).
59. On the role of Schlick, see Engler and Renn (2013).
60. Kretschmann (1917). For Einstein's reaction, see Einstein (1918b). For historical discussion, see Rynasiewicz (1999), see also Gutfreund and Renn (2017).
61. See *The Collected Papers of Albert Einstein*, available online at http://einsteinpapers.press.princeton.edu/. See in particular "The Einstein–de Sitter–Weyl–Klein Debate," in *CPAE* 8A, 351–57. See also Renn (2007d); Janssen (2014b).
62. Friedmann (1922), Lemaître (1927). See the discussion in Eisenstaedt (1993), and also in Gutfreund and Renn (2017), 50–51.
63. Einstein (1922).
64. Hubble (1929).
65. See Einstein (1931) and Gutfreund and Renn (2015b, 153–54; 2017, ch. 5).
66. Gutfreund and Renn (2017).
67. For an analysis of Schwarzschild's work, see Eisenstaedt (1982, 1989a) and Schemmel (2005). For a general account of these years, see Gutfreund and Renn (2017, 46–47).
68. For historical discussion, see Kennefick (2012).
69. Kragh (1996).
70. Janssen (2014b).
71. See Renn (2007c). For a comprehensive history of gravitational waves, see Kennefick (2007).
72. Abraham (1913).
73. "Discussion Following Lecture Version of 'On the Present State of the Problem of Gravitation,'" in *CPAE* 4E, Doc. 18, 229.
74. See Buchwald, Gutfreund, and Renn (2016).

75. Karl Schwarzschild to Arnold Sommerfeld, February 17, 1916, Archiv Deutsches Museum München, NL 89, 059. I am grateful to Matthias Schemmel for bringing this letter to my attention.
76. Einstein to Karl Schwarzschild, February 19, 1916, *CPAE* 8E, Doc. 194, 196.
77. Einstein to Karl Schwarzschild, February 19, 1916, *CPAE* 8E, Doc. 194, 196.
78. The letter is lost; the reply is Einstein to Willem De Sitter, June 22, 1916, *CPAE* 8E, Doc. 227, 223.
79. Einstein (1916a).
80. Kennefick (2007, ch. 3). See also Janssen and Renn (forthcoming, Part I, Section 7.2).
81. Einstein (1918a), in *CPAE* 7E, Doc. 1, 10.
82. See Kennefick (2007).
83. For the following, see Kennefick (2005).
84. Einstein to John Tate, July 27, 1936, EA 19-086. For early refereeing practices in the *Physical Review*, see Lalli (2016).
85. Einstein and Rosen (1937).
86. See, e.g., Rosen (1937, 1956). On Infeld's attitude, see Infeld and Plebanski (1960).
87. Eisenstaedt (1986, 1989b, 2006).
88. Blum, Lalli, and Renn (2015).
89. Blum, Lalli, and Renn (2018).
90. Will (1989). For recent reassessments of the renaissance, see Blum, Lalli, and Renn (2015, 2016, 2018); Blum et al. (2017b), see also Lalli (2017). The following is based on collaboration with Alexander Blum and Roberto Lalli.
91. Thorne (1994). See also Goenner (2017).
92. The formation of the relativity community is analyzed extensively in a pioneering study by Roberto Lalli (2017).
93. See Kaiser (2005, 2012).
94. Mercier and Kervaire (1956).
95. Blum, Lalli, and Renn (2015), 613; see also Schutz (2012).
96. See, e.g., Bergmann (1956).
97. Of crucial influence were the discussions at the Chapel Hill conference, involving among others Richard Feynman, John Wheeler, and Felix Pirani (DeWitt and Rickles 2011). For historical discussion on which the following is based, see Collins (2004); Trimble (2017); Kennefick (2007). For a comprehensive survey of experimental tests of general relativity, see Will (1986).
98. Weber (1969).
99. For reviews, see Choptuik, Lehner, and Pretorius (2015), Kramer (2016). For the historical context, see the discussion in Blum, Lalli and Renn (2018).
100. Hulse and Taylor (1975); Taylor and Weisberg (1982). For a historical discussion, see Kennefick (2017).

Collins: The Detection of Gravitational Waves

1. The study of gravitational wave detection physics has been the backbone of my fieldwork over my career and has led to many spin-offs, such as contributions to the philosophical and sociological understanding of science and the analysis of expertise.
2. By "direct" observation I mean a method in which the interaction of the gravitational waves with the observing instruments gives rise to the signal. In the case of Taylor and Weisberg, it was the *electromagnetic* radiation from the binary pulsar that interacted with their observing instrument (Taylor and Weisberg 1982; Weisberg and Taylor 2004). Taylor and Hulse, were awarded the Nobel Prize but this award was for the discovery of the binary pulsar. As I understand it, Damour worked out the expected rate of decay of the binary pulsar that would accord with Einstein's quadrupole formula for the emission of gravitational waves from such

a system, and Weisberg did most of the data collection and analysis. Though Taylor and his colleagues believe this was a "direct" observation of gravitational waves, under my definition (which fits with general opinion among gravitational wave physicists), this could only be true if we construe the binary pulsar itself as an "observing instrument," which seems to stretch normal usage since no one constructed the binary pulsar and no one can make any more of them in order to, say, speed up the rate of observation. For the direct/indirect controversy, see Collins (2017).

3. Bartusiak (2000) has an account of the history of the space antenna efforts.
4. Thanks to Jim Hough and Peter Bender.
5. Of course, string theory, it is claimed by some, has too much reality in physics departments, books, and papers, considering that it is not matched by experiment or observation, so that is an instance of the bearing of this sociological view of the world on narrow questions of theory versus experiment and observation.
6. There is no deeply *philosophical* claim here about the reality of germs; the argument is about the felt reality of germs and other things, not their ontological status. Remember, there are many things in the world that are as yet undiscovered and, in the sense in which "reality" is used here, they have no reality for us irrespective of any causal impact they may have on us since that causal impact is unknown to us, while other things that we normally take to be undiscoverable, such as witches and mermaids, can have a felt reality.
7. Weber said to me in 1972:

   I don't think I've ever made the statement that I've discovered gravitational radiation. . . . So the position that I have taken from the start—my first paper was entitled "Evidence for the Discovery of Gravitational Radiation"—and I've always taken the position that I do experiments and I gather evidence, and I've always taken the position that the evidence suggests gravitational radiation has been discovered but I've never said publicly or privately, or written, that I've discovered anything.
8. The immediate acceptance by the scientific community was very strange. Partly it was due to the convincing look of the superimposed waveforms from the two sites, but no outsiders really understood how, for instance, the statistical significance was arrived at. The almost uniform acceptance of the result without questioning must have to do with the fact that all the experts were inside the collaboration, and also reflected the history of skepticism about previous supposed discoveries and the sheer amount of time and effort that had been put into perfecting the instruments. In a deep way, the acceptance of any scientific finding is a matter of trust, but here it was trust in a shallow way, too.
9. People usually say one-thousandth of the diameter of a proton, but that is peak-to-trough of the wave, whereas one has to measure much smaller things if the shape is to be revealed.
10. Thorne, of course, was also one of the chief theorists and made contributions to the experiment.
11. Drever took up a part-time post at Caltech in 1979 and shifted to full-time in 1983.
12. My 2004 footnote gives too much credit to Drever for "inventing the use of the Fabry-Perot cavity" for long interferometers, as it had already been used in a 30-meter, single-armed interferometer at JILA. I am not sure whether the control mechanism that he brought to LIGO was already in place in that device. (Thanks to Jim Hough for drawing the JILA device to my attention and to Robin "Tuck" Stebbins for giving me the details.)

Kormos Buchwald: Einstein at Caltech

1. Judith R. Goodstein, *Millikan's School,* New York: W. W. Norton, 1991.
2. *CPAE* Vol. 2, Doc. 14.
3. *CPAE* Vol. 5, Docs. 56 and 60.
4. At the time of the letter, the United States and Germany were still officially at war. Einstein to Carl Beck, April 8, 1921, *CPAE* Vol. 12, Doc. 115.

5. Einstein to Ludwik Silberstein, Berlin, October 4, 1921, *CPAE* Vol. 12, Doc. 254.
6. See Millikan to Einstein, May 22, 1922, *CPAE* Vol. 13, Doc. 199.
7. The equivalent of $9,000 per month in 2019.
8. Millikan to Ehrenfest, June 22, 1922, Albert Einstein Archives [94 985]. Pirenne had visited several American universities in late 1922. He was hosted by Millikan for three days in Pasadena and gave a lecture at Caltech on December 4, 1922 (see Sarah Keymeulen and Jo Tollebeek, *Henri Pirenne, Historian: A Life in Pictures,* Lipsius Leuven: Leuven University Press, 2011). It does not appear that he returned to Caltech for a longer visit.
9. Millikan to Hale, September 29, 1922, in Nathan Reingold and Ida H. Reingold, *Science in America: A Documentary History 1900–1939,* Chicago and London: University of Chicago Press, p. 365.
10. Nicholas M. Butler and Gano Dunn to Einstein, February 26, 1923 and Gano Dunn to Einstein, March 1, 1923 (*CPAE* Vol. 13, Docs. 432 and 435); and Einstein to Gano Dunn, April 11, 1923, *CPAE* Vol. 14, Doc. 11.
11. Millikan to Einstein, November 27, 1923, *CPAE* Vol. 14, Doc. 162.
12. Ehrenfest to Einstein, May 25, 1924, *CPAE* Vol. 14, Doc. 255.
13. Einstein to Ehrenfest, May 31, 1924, *CPAE* Vol. 14, Doc. 259.
14. Robert A. Millikan to Einstein, October 2, 1924, *CPAE* Vol. 14, Doc. 329.
15. For this exchange, see *CPAE* Vol. 14, Docs. 359, 360, 390, and 400.
16. Einstein to Edwin E. Slosson, between June 26, 1925 and July 31, 1925, *CPAE* Vol. 15, Doc. 13.
17. Epstein to Einstein, July 25, 1925, *CPAE* Vol. 15, Doc. 31.
18. Einstein to Ehrenfest, August 18, 1925, *CPAE* Vol. 15, Doc. 49.
19. Einstein to Millikan, September 1, 1925, *CPAE* Vol. 15, Doc. 58.
20. "Einstein Is Ready to Bet on His Theory," *San Francisco Examiner,* January 16, 1926, p. 6 (*CPAE* Vol. 15, Doc. 160).
21. Roy J. Kennedy, "A Refinement of the Michelson-Morley Experiment." *National Academy of Sciences: Publications* 12 (1926): 621–629. See also the Introduction to *CPAE* Vol. 15 and Roberto Lalli, "The Reception of Miller's Ether-Drift Experiments in the USA," *Annals of Science* 69 (2012): 153–214.
22. Epstein to Einstein, September 28, 1926, *CPAE* Vol. 15, Doc. 372; Millikan to Einstein, December 3, 1926, *CPAE* Vol. 15, Abs. 662.
23. See the Introduction to *CPAE* Vol. 14, p. xxxvi.
24. George E. Hale to Einstein, December 2, 1926, *CPAE* Vol. 15, Doc. 425.
25. Einstein to Hale, December 25, 1926, *CPAE* Vol. 15, Doc. 437.
26. Veblen was writing from Hampstead, London, where he was residing with his wife's brother, O. W. Richardson. Oswald Veblen to Einstein, September 16, 1927, AEA 23-149.
27. Einstein to O. Veblen, September 17, 1927, AEA 23-150.
28. For Veblen's role a few years later in the establishment of the Institute for Advanced Study in Princeton, and the negotiations between Abraham Flexner, its founding director, and various scientists, including Einstein, see Steve Batterson, *Pursuit of Genius: Flexner, Einstein, and the Early Faculty at the Institute for Advanced Study,* Wellesley, MA: A. K. Peters, 2006.
29. Millikan to Einstein, August 25, 1929, AEA 17-303.
30. Einstein to Millikan, September 18, 1929 (received October 7, 1929), AEA 17-304; Elsa Einstein to Millikan, September 18, 1929, AEA 17-305.
31. The first known prediction of this kind appears in Einstein's paper, "On the Relativity Principle and the Conclusions Drawn from It," conceived in late 1907 (see *CPAE* Vol. 2, Doc. 47).
32. The work was carried out in 1919 and 1920, but not all analysts were convinced that evidence for a redshift had been found. See *CPAE* Vol. 7, Doc. 31, especially note 47, and Klaus Hentschel, "Grebe/Bachems photometrische Analyse der Linienprofile und die Gravitations-Rotverschiebung: 1919 bis 1922," *Annals of Science* 49 (1992): 21–46.
33. Allan Sandage, *Centennial History of the Carnegie Institution of Washington: Volume 1, The Mount Wilson Observatory*. Cambridge: Cambridge University Press, 2005, p. 137.

34. Sandage, *Centennial History*, pp. 500 and 501.
35. "Kosmologische Betrachtungen zur allgemeinen Relativitätstheorie," *Königlich Preußische Akademie der Wissenschaften (Berlin). Sitzungsberichte* (1917): 142–152. See *CPAE* Vol. 6, Doc. 43.
36. "Zum kosmologischen Problem der allgemeinen Relativitätstheorie," *Preussische Akademie der Wissenschaften, Sitzungsberichte. Phys.-Math Klasse. Sonderausgabe* (1931) XII: 3–5 (issued May 9, 1931). For an interpretation and translation of the paper into English, see C. O'Raifeartaigh and B. McCann, "Einstein's Cosmic Model of 1931 Revisited: An Analysis and Translation of a Forgotten Model of the Universe," *European Physical Journal H* (2014) 39: 63–85. See also Robert Fox, "Einstein in Oxford," *Notes and Records. The Royal Society Journal of the History of Science* (2019) 72: 293–318, and the references therein.
37. The *New York Times*, January 3, 1931. A draft translation of Einstein's answers into English in Elsa Einstein's hand is available AEA 23-468. Milton Humason began work as a night janitor at the Mount Wilson observatory and then became a talented observer, assisting Edwin Hubble. Richard C. Tolman, professor at Caltech, was an early researcher into Einstein's theory of relativity and its applications to cosmology.
38. Georges Lemaître, "Rencontres avec A. Einstein," *Revue des questions scientifiques* 129 (1958): 129–132. Texte lu à la radio nationale belge le 27 avril 1957 en commémoration du deuxième anniversaire de la mort d'Einstein : "J'ai rencontré Einstein pour la première fois, il y a vingt-neuf ans. Il était venu à Bruxelles assister au congrès Solvay de 1927. En se promenant dans les allées du parc Leopold, il me parla d'un article, peu remarqué, que j'avais écrit l'année précédente sur l'expansion de l'univers et qu'un ami lui avait fait lire. Après quelques remarques techniques favorables, il conclut en disant que du point de vue physique cela lui paraissait tout à fait abominable. . . . Dans le taxi, je parlai des vitesses des nébuleuses et j'eus l'impression qu'Einstein n'était guère au courant des faits astronomiques."
39. Approximately equivalent to $23,000 to $31,000 in 2019.
40. *New York Times*, December 17, 1930.
41. Einstein's assistant, Helen Dukas, later transcribed and typed the diaries' text, and inserted further details and explanations.
42. Entry of Monday, December 14, 1930 (Einstein erroneously entered December 15), Einstein Travel Diary Pasadena, Albert Einstein Archives 29–134.
43. Einstein travel diary to the United States, entries of January 1 and 3, 1931, and Helen Dukas comments in its transcribed version.
44. Einstein to Millikan, August 1, 1931, AEA 90-317. Here, as in other instances, Einstein misspells Hubble's name as "Hubbel."
45. Approximately equivalent to $350,000 in 2019. AEA 90-409.
46. See, e.g., George E. Hale to Arthur A. Noyes, October 4, 1931, Caltech Institute Archives.
47. Elsa Einstein to Millikan, October 9, 1931. AEA 17-314.
48. Millikan wrote to Einstein October 11, 1931, while traveling from Vienna to Rome. AEA 17-315.
49. "Einheitliche Theorie von Gravitation und Elektrizität," presented to the Prussian Academy of Sciences on October 22, 1931, published December 2, 1931.
50. Einstein to Millikan, October 19, 1931. AEA 17-313.
51. AEA 90-411 and Elsa Einstein to Millikan, November 14, 1931. Elsa wrote that she would try to rent the house on S. Oakland again, and that a secretary for three times a week or a few hours each would be sufficient. AEA 17-318.
52. Millikan to Einstein, April 14, 1932, AEA 17-323.
53. Elsa Einstein to Millikan, August 13, 1932, AEA 17-330. In fact, Mayer's appointment at the Institute for Advanced Study was to be rather precarious. For a detailed discussion, see Batterson, *Pursuit of Genius*, pp. 116–17.
54. Elsa Einstein to Veblen, 18 August 1932, AEA 23-160.
55. Millikan to Einstein, August 15, 1932, AEA 17-331.
56. Kultusministerium to Prussian Academy of Sciences, August 31, 1932, Geheimes Staatsarchiv Berlin, AEA 83-368.
57. Einstein to Prussian Academy of Sciences, September 5, 1932, AEA 29-284.

58. Veblen had informed Einstein of being elected to the Nassau Club in Princeton and invited him to speak to the Mathematical Society in December 1933. Einstein to Veblen, February 14, 1933, AEA 23-163.
59. Millikan to Einstein, February 23, 1933, AEA 17-346.
60. First published on March 11, 1933 in *New York World Telegram*. For drafts of this declaration, see AEA 28-236 and 28-236.1. Elsa Einstein wrote on the verso: "Wichtig! [important] statements." AEA 28-237.

Howard: How General Relativity Shaped Twentieth-Century Philosophy of Science

* Correspondence that has already been published in The Collected Papers of Albert Einstein (Princeton, NJ: Princeton University Press, 1987–present) is cited by volume number, document number, and page, after the model "CPAE-x, Doc. yyy, z." Unpublished items from Einstein's correspondence are cited by their control index numbers in the Einstein Archive, after the model "EA xx-xxx."

1. Einstein probably read the 1895 German translation by Theodore Lipps, the first volume of which is titled *Ein Traktat über die menschliche Natur: Ein Versuch die Methode der Erfahrung in die Geisteswissenschaften einzuführen*. Vol. 1. *Üben den Verstand* (Hume 1895). In his December 1915 letter to Mortiz Schlick, Einstein refers to his having studied Hume's "Traktat über den Verstand," the mentioned title making sense only if it were the Lipps translation that Einstein had in mind. See Einstein to Schlick, December 14, 1915 (*CPAE*-8, Doc. 165, 220).
2. Probably either the 1877 German translation by Schiele (Hume 1877) or the 1884–1886 translation by Gomperz (Hume 1884–1887).
3. Perhaps the 1900 second edition. No German translation was available.
4. Probably the 1904 German translation by Lindemann (Poincaré 1904).
5. There is a large and helpful literature on the history of Einstein and the discovery of general relativity, starting with the pioneering papers by John Stachel (1980) and John Norton (1984) and culminating in the definitive history by Jürgen Renn and colleagues (Renn et al. 2007). A recent, accessible summary of the story is to be found in Janssen and Renn 2015.
6. Schneider, later Rosenthal-Schneider, published a useful memoir of her discussions with Einstein (Rosenthal-Schneider 1980).
7. Reichenbach's snide remark about the "epigones of the great Kant" alludes to the title of the book that did more than any other to call into being the neo-Kantian revival of the latter nineteenth century as a self-aware philosophical movement, Otto Liebmann's *Kant und die Epigonen* (Liebmann 1865). Liebmann's book had been republished on the occasion of his death in 1912, Carnap's future, Jena dissertation director, Bauch, being the editor.
8. Different thinkers have defended one or another version of the doctrine of the contingent a priori. Best known today is Michael Friedman's revival of Reichenbach's specific project for the purpose of promoting a neo-Kantian counter-narrative to the dominant, twentieth-century, empiricist narrative in the philosophy of science. See Friedman (2001) and Howard (2010).
9. The Schlick-Reichenbach correspondence is contained in the Hans Reichenbach papers in the Archive for Scientific Philosophy at the University of Pittsburgh. Items in the correspondence are cited by their control index numbers after the model HR-xxxx-xx-xx.
10. The changes in the second edition and the shift in Schlick's views about the distinction between definitions and empirical judgments is discussed in considerable detail in Howard (1994).
11. See Howard (1996) for a detailed study of Carnap's extensive efforts to disentangle and properly appreciate the import of different notions of completeness.
12. When, in 1986, I asked Quine whether he knew of these remarks by Einstein at the time when he was writing "Two Dogmas of Empiricism," he responded: "When I wrote 'Two Dogmas of Empiricism,' I had not read Einstein's reply to Reichenbach. My holism there was just my

own common sense, plus perhaps some influence from Neurath's congenial figure of the boat. After 'Two Dogmas' appeared, January 1951, both Hempel and Philipp Frank told me about the kinship of my view to Duhem's; so I added the footnote citation of Duhem when 'Two Dogmas' was reprinted in *From a Logical Point of View*, 1953" (W.V.O. Quine, private communication).

References

Barish: The Quest for (and Discovery of) Gravitational Waves

Abbott, B. P. et al. 2016a. (LIGO Scientific Collaboration and Virgo Collaboration). "Observation of Gravitational Waves from a Binary Black Hole Merger." *Physical Review Letters* 116 (24): 061102.

Abbott, B. P. et al. 2016b. (LIGO Scientific Collaboration and Virgo Collaboration). "Observation of Gravitational Waves from a 22-Solar-Mass Binary Black Hole Coalescence." *Physical Review Letters* 116 (24): 241103.

Thorne: One Hundred Years of Relativity: From the Big Bang to Black Holes and Gravitational Waves

Hawking, S., K. S. Thorne, I. Novikov, T. Ferris, and A. Lightman. 2002. *The Future of Spacetime,* edited by R. H. Price. New York: Norton.

James, O., E. Von Tunzelmann, P. Franklin, and K. S. Thorne. 2015. "Gravitational lensing by spinning black holes in astrophysics and in the movie Interstellar," *Classical and Quantum Gravity* 32, 0650001.

Kennefick, D. 2007. *Traveling at the Speed of Thought: Einstein and the Quest for Gravitational Waves.* Princeton, NJ: Princeton University Press.

Thorne, K. S. 2014. *The Science of Interstellar.* New York: Norton.

Thorne, K. S. 2019. Biographical and Nobel Lecture. https://www.nobelprize.org/prizes/physics/2017/thorne/facts/

Buonanno: The New Era of Gravitational-Wave Physics and Astrophysics

Abbott, B. P. et al. 2016a. (LIGO Scientific Collaboration and Virgo). *Physical Review Letters* 116, 061102, arXiv:1602.03837 [gr-qc].

Abbott, B. P. et al. *2016b.* (LIGO Scientific Collaboration and Virgo). *Physical Review Letters* 116, 241103, arXiv:1606.04855 [gr-qc].

Abbott, B. P. et al. *2016c.* (LIGO Scientific Collaboration and Virgo). *Physical Review Letters* X6, 041015, arXiv:1606.04856 [gr-qc].

Abbott, B. P. et al. 2016d. (LIGO Scientific Collaboration and Virgo). *Physical Review Letters* 116, 241102, arXiv:1602.03840 [gr-qc].

Abbott, B. P. et al. 2016e. (LIGO Scientific Collaboration and Virgo). *Astrophysical Journal* 818, L22, arXiv:1602.03846 [astro-ph.HE].

Abbott, B. P. et al. 2016f. (LIGO Scientific Collaboration and Virgo). , arXiv:1602.03841 [gr-qc].

Armano, M., et al. 2016. *Physical Review Letters* 116, 231101.

Buonanno, A., and B. S. Sathyaprakash. 2015. In *General Relativity and Gravitation: A Centennial Perspective,* edited by A. Ashtekar et al. Cambridge University Press, 287–346. arXiv:1410.7832 [gr-qc].

Choptuik, M. W., L. Lehner, and F. Pretorius. 2015. In *General Relativity and Gravitation: A Centennial Perspective*, edited by A. Ashtekar et al. Cambridge University Press, 361–401. arXiv:1502.06853 [gr-qc].

Kennefick: The Wagers of Science

Abbott, B. P. et al. 2016. (LIGO Scientific Collaboration and Virgo Collaboration). "Observation of Gravitational Waves from a Binary Black Hole Merger." *Physical Review Letters* 116, 061102.

Anonymous. 2016. *A Brief History of Ligo*. Available, as of 3/31/2017, at https://www.ligo.caltech.edu/system/media_files/binaries/313/original/LIGOHistory.pdf.

Baker, J. G., Joan Centrella, D. I. Choi, M. Koppitz, and J. van Meter. 2006. "Gravitational-Wave Extraction from an Inspiraling Configuration of Merging Black Holes." *Physical Review Letters* 96, 111102.

Bartusiak, Marcia. 2017. *Einstein's Unfinished Symphony: The Story of a Gamble, Two Black Holes, and a New Age of Astronomy*. New Haven, CT: Yale University Press. This book is an updated version of *Einstein's Unfinished Symphony: Listening to the Sounds of Space-Time* (Washington, DC: Joseph Henry Press, 2000).

Bondi, Hermann. 1957. "Plane Gravitational Waves in General Relativity." *Nature* 179, 1072–1073.

Bondi, Hermann, Felix A. E. Pirani, and Ivor Robinson. 1959. "Gravitational Waves in General Relativity III. Exact Plane Waves." *Proceedings of the Royal Society of London, series A* 251, 519–533.

Bondi, Hermann, M.G.J. van der Burg, and A.W.K. Metzner. 1962. "Gravitational Waves in General Relativity VII: Waves from Axi-Symmetric Isolated Systems." *Proceedings of the Royal Society of London, series A* 269, 21–52.

Campanelli, Manuella, Carlos Lousto, Pedro Marronetti, and Yosef Zlochower. 2006. "Accurate Evolutions of Orbiting Black-Hole Binaries without Excision." *Physical Review Letters* 96, 111101.

Cardoso, Vitor, Leonardo Gualtieri, Carlos A. R. Herdeiro, and Ulrich Sperhake. 2015. "Exploring New Physics Frontiers Through Numerical Relativity." *Living Reviews in Relativity* 18, 1.

Collins, Harry M. 2004. *Gravity's Shadow: The Search for Gravitational Waves*. Chicago: University of Chicago Press.

Collins, Harry M. 2014. *Gravity's Ghost and Big Dog: Scientific Discovery and Social Analysis in the Twenty-First Century*. Chicago: University of Chicago Press.

Collins, Harry M. (2017). *Gravity's Kiss: The Detection of Gravitational Waves*. Cambridge, MA: MIT Press.

Cutler, Curt, Theocharis A. Apostolatos, Lars Bildsten, Lee Samuel Finn, Eanna E. Flanagan, Daniel Kennefick, Draza M. Markovic, Amos Ori, Eric Poisson, Gerald J. Sussman, and Kip S. Thorne. 1993. "The Last Three Minutes: Issues in Gravitational-Wave Measurements of Coalescing Compact Binaries." *Physical Review Letters* 70, 2984–87.

De Witt, Cecile M. 1957. *Conference on the Role of Gravitation in Physics* (WADC Technical Report 57-216, Wright-Patterson Air Force Base, Ohio). This hard-to-find report has been republished in electronic form, with supporting documents, as De Witt, Cecile M. and Dean Rickles. 2011. *The Role of Gravitation in Physics— Report from the 1957 Chapel Hill Conference* (Edition Open Access). It is available on the web at http://www.edition-open-sources.org/sources/5/index.html. In addition, Volume 29, Issue 3 of *Reviews of Modern Physics* was devoted almost exclusively to topics in gravitational physics, with a number of the papers related to presentations given at the conference. A summary of the conference was given in that issue by Bergmann, Peter. 1957. "Summary of the Chapel Hill Conference." *Reviews of Modern Physics* 29, 352–354.

Eddington, Arthur Stanley. 1922. "The Propagation of Gravitational Waves." *Proceedings of the Royal Society of London, series A* 102, 268–82.

Einstein, Albert. 1916. "Naherungsweise Integration der Feldgleichungen der Gravitation." *Koniglich Preussische Akademie der Wissenschaften Berlin, Sitzungsberichte*, 688–96. Reprinted as Doc. 32 of Vol. 6 of the *Collected Papers of Albert Einstein*. Princeton, NJ: Princeton University Press, 1997.

Einstein, Albert. 1918. “Uber Gravitationswellen.” *Koniglich Preussische Akademie der Wissenschaften Berlin, Sitzungsberichte*, 154–67. Reprinted as Doc. 1 of Vol. 7 of the *Collected Papers of Albert Einstein*. Princeton, NJ: Princeton University Press, 2002.

Einstein, Albert. 1998. *Collected Papers of Albert Einstein*, Volume 8. Princeton, NJ: Princeton University Press.

Einstein, Albert, and Nathan Rosen. 1937. “On Gravitational Waves.” *Journal of the Franklin Institute* 223, 43–54.

Hahn, Susan G., and Richard W. Lindquist. 1964. “The Two-Body Problem in Geometrodynamics.” *Annals of Physics* 29, 304–31.

Holst, Michael, Olivier Sarbach, Manuel Tiglio, and Michele Vallisneri. 2016. “The Emergence of Gravitational Wave Science: 100 Years of Development of Mathematical Theory, Detectors, Numerical Algorithms, and Data Analysis Tools.” *Bulletin (New Series) of the American Mathematical Society* 53, 513–54.

Hulse, Russell A., and Joseph H. Taylor. 1975. “Discovery of a Pulsar in a Binary System.” *Astrophysical Journal* 195, L51–L53.

Infeld, Leopold. 1941. *Quest— The Evolution of a Physicist.* London: Gollancz.

Kennefick, Daniel. 2005. “Einstein versus the Physical Review.” *Physics Today* 58, 43–48. This article is available online at http://physicstoday.scitation.org/doi/full/10.1063/1.2117822.

Kennefick, Daniel. 2007. *Traveling at the Speed of Thought: Einstein and the Quest for Gravitational Waves.* Princeton, NJ: Princeton University Press.

Lehner, Luis. 2001. “Numerical Relativity: A Review.” *Classical and Quantum Gravity* 18, 101345.

Levin, Janna. 2016. *Black Hole Blues and Other Songs from Outer Space.* New York: Alfred A. Knopf.

Pretorius, Frans. 2005. “Evolution of Binary Black-Hole Spacetimes.” *Physical Review Letters* 95, 121101.

Rosen, Nathan. 1955. “On Cylindrical Gravitational Waves.” *Jubilee of Relativity Theory Proceedings of the Anniversary Conference at Bern, July 11–16, 1955.* Basel: Birkhauser-Verlag, 171–75.

Smarr, Larry. 1977. “Space-Times Generated by Computers: Black Holes with Gravitational Radiation.” *Annals of the New York Academy of Sciences* 302, 569–604.

Sperhake, Ulrich. 2015. “The Numerical Relativity Breakthrough for Binary Black Holes.” *Classical and Quantum Gravity* 32, 124011.

Thorne, Kip S. 1995. *Black Holes and Time Warps: Einstein's Outrageous Legacy.* New York: Norton.

Weber, Joseph. 1969. “Evidence for Discovery of Gravitational Radiation.” *Physical Review Letters* 22, 1320–24.

Weinberg, Steven. 1993. *The First Three Minutes: A Modern View of the Origin of the Universe.* New York: Basic Books.

Weisberg, Joel M., and Joseph H. Taylor. 1981. “Gravitational Radiation from an Orbiting Pulsar.” *General Relativity and Gravitation* 13, 1–6.

Renn: The Genesis and Transformation of General Relativity

Abbott, B. P. et al. 2016. “Observation of Gravitational Waves from a Binary Black Hole Merger.” *Physical Review Letters* 116, 061102.

Abraham, Max. 2013. “Eine neue Gravitationstheorie.” *Archiv der Mathematik und Physik. Third series* 20: 193–209. English translation: “A New Theory of Gravitation,” in *Gravitation in the Twilight of Classical Physics: Between Mechanics, Field Theory, and Astronomy*, edited by J. Renn and M. Schemmel, *The Genesis of General Relativity*, vol. 3 (Dordrecht: Springer), 347–62.

Bergmann, P. G. 1956. “Fifty Years of Relativity.” *Science* 123: 486–94.

Blum, Alexander S., Domenico Giulini, Roberto Lalli, and Jürgen Renn, eds. 2017a. “The Renaissance of Einstein's Theory of Gravitation.” *Special Issue of European Physical Journal H* 42(2).

Blum, Alexander S., Domenico Giulini, Roberto Lalli, and Jürgen Renn. 2017b. “Editorial Introduction to the Special Issue ‘The Renaissance of Einstein's Theory of Gravitation.’” *European Physical Journal H* 42(2): 95–105 (special issue).

Blum, Alexander S., Roberto Lalli, and Jürgen Renn. 2015. "The Reinvention of General Relativity: A Historiographical Framework for Assessing One Hundred Years of Curved Space-time." *Isis* 106, 3: 598–620.

Blum, Alexander S., Roberto Lalli, and Jürgen Renn. 2016. "The Renaissance of General Relativity: How and Why It Happened." *Annals of Physics* 528(5): 344–49.

Blum, Alexander S., Roberto Lalli, and Jürgen Renn. 2018. "Gravitational Waves and the Long Relativity Revolution." *Nature Astronomy* 2: 534–43.

Blum, Alexander S., Jürgen Renn, Donald C. Salisbury, Matthias Schemmel, and Kurt Sundermeyer. 2012. "1912: A Turning Point on Einstein's Way to General Relativity." *Annalen der Physik* 1: A11–A13.

Born, Max. 1914. "Der Impuls-Energie-Satz in der Elektrodynamik von Gustav Mie." *Nachrichten von der Königlichen Gesellschaft der Wissenschaften zu Göttingen* 1: 23–36. English translation: "The Momentum-Energy Law in the Electrodynamics of Gustav Mie." In *From an Electromagnetic Theory of Matter to a New Theory of Gravitation*, edited by J. Renn and M. Schemmel. *The Genesis of General Relativity*, vol. 4 (Dordrecht: Springer, 2007), 745–56.

Buchwald, Diana, Hanoch Gutfreund, and Jürgen Renn. 2016. "100 Jahre Gravitationswellen: Kräuselungen der Raumzeit verloren und wiedergefunden." *Frankfurter Allgemeine*, February 17.

Buchwald, Diana K., Jürgen Renn, and Robert Schlögl. 2013. "A Note on Einstein's Scratch Notebook of 1910–1913." In *Physics as a Calling, Science for Society: Studies in Honour of A. J. Kox*, edited by A. Maas and H. Schatz (Leiden: Leiden University Press), 81–88.

Choptuik, Matthew W., Louis Lehner, and Frans Pretorius. 2015. "Probing Strong Field Gravity Through Numerical Simulations." In *General Relativity and Gravitation: A Centennial Perspective*, edited by A. Ashtekar et al. (Cambridge: Cambridge University Press, 2015), 361–411.

Collins, Harry. 2004. *Gravity's Shadow: The Search for Gravitational Waves*. Chicago: University of Chicago Press.

Corry, Leo, Jürgen Renn, and John Stachel. 1997. "Belated Decision in the Hilbert-Einstein Priority Dispute." *Science* 278: 1270–73.

CPAE 2E. 1989. *The Collected Papers of Albert Einstein. The Swiss Years: Writings 1900–1909* (English translation supplement), vol. 2 (Princeton, NJ: Princeton University Press).

CPAE 3. 1994. *The Collected Papers of Albert Einstein. The Swiss Years: Writings 1909–1911*, vol. 3 (Princeton, NJ: Princeton University Press).

CPAE 4. 1996. *The Collected Papers of Albert Einstein. The Swiss Years: Writings 1912–1914*, vol. 4 (Princeton, NJ: Princeton University Press).

CPAE 4E. 1996. *The Collected Papers of Albert Einstein. The Swiss Years: Writings 1912–1914 (English translation supplement)*, vol. 4 (Princeton, NJ: Princeton University Press).

CPAE 5E. 1995. *The Collected Papers of Albert Einstein. The Swiss Years: Correspondence, 1902–1914 (English translation supplement)*, vol. 5 (Princeton, NJ: Princeton University Press).

CPAE 6E. 1997. *The Collected Papers of Albert Einstein. The Berlin Years: Writings 1914–1917 (English translation supplement)*, vol. 6 (Princeton, NJ: Princeton University Press).

CPAE 7E. 2002. *The Collected Papers of Albert Einstein: The Berlin Years: Writings 1918–1921* (English translation supplement), vol. 7 (Princeton, NJ: Princeton University Press).

CPAE 8A. 1998. *The Collected Papers of Albert Einstein: The Berlin Years: Correspondence, 1914–17*, vol. 8, part A (Princeton: Princeton University Press).

CPAE 8E. 2002. *The Collected Papers of Albert Einstein: The Berlin Years: Correspondence, 1914–18* (English translation supplement), vol. 8 (Princeton, NJ: Princeton University Press).

DeWitt, Cécile M., and Dean Rickles, eds. 2011. *The Role of Gravitation in Physics. Report from the 1957 Chapel Hill Conference*. Sources 5: Max Planck Research Library for the History and Development of Knowledge (Berlin: Edition Open Access).

Droste, Johannes. 1914. "On the Field of a Single Centre in Einstein's Theory of Gravitation." *Koninklijke Akademie van Wetenschappen te Amsterdam, Section of Sciences Proceedings* 17: 998–1011.

Earman, John, and Michel Janssen. 1993. "Einstein's Explanation of the Motion of Mercury's Perihelion." In *The Attraction of Gravitation*, Einstein Studies vol. 5, edited by J. Earman, M. Janssen, and J. D. Norton (Boston: Birkhäuser), 129–72.

Einstein, Albert. 1907. "On the Relativity Principle and the Conclusions Drawn from It." In *CPAE* 2E, Doc. 47, 252–311.

Einstein, Albert. 1914. "The Formal Foundation of the General Theory of Relativity." In *CPAE* 6E, Doc. 9, 30–84.

Einstein, Albert. 1915. "Explanation of the Perihelion Motion of Mercury from the General Theory of Relativity." In *CPAE* 6E, Doc. 24, 112–16.

Einstein, Albert. 1916a. "Approximative Integration of the Field Equations of Gravitation." In *CPAE* 6E, Doc. 32, 201–210.

Einstein, Albert. 1916b. "Hamilton's Principle and the General Theory of Relativity." In *CPAE* 6E, Doc. 41, 240–246.

Einstein, Albert. 1916c. "The Foundation of the General Theory of Relativity." In *CPAE* 6E, Doc. 30, 147–200.

Einstein, Albert. 1918a. "On Gravitational Waves." In *CPAE* 7E, Doc. 1, 9–27.

Einstein, Albert. 1918b. "On the Foundations of the General Theory of Relativity." In *CPAE* 7E, Doc. 4, 33–35.

Einstein, Albert. After 1920. "Fundamental Ideas and Methods of the Theory of Relativity, Presented in Their Development." in *CPAE* 7E, Doc. 31, 113–50.

Einstein, Albert. 1922. "Comment on A. Friedmann's Paper: 'On the Curvature of Space.'" In *CPAE* 13E, Doc. 340, 271–72.

Einstein, Albert. 1931. "Zum kosmologischen Problem der allgemeinen Relativitätstheorie." *Sitzungsberichte der Preußische Akademie der Wissenschaften*, 235–37.

Einstein, Albert. 1936. "Physik und Realität." *Journal of the Franklin Institute* 221: 313–17.

Einstein, Albert. [1949] 1979. *Autobiographical Notes*, edited by P. A. Schilpp (La Salle, IL: Open Court).

Einstein, Albert. 1954. *Ideas and Opinions* (New York: Crown).

Einstein, Albert, and Marcel Grossmann. 1913. "Outline of a Generalized Theory of Relativity and of a Theory of Gravitation." In *CPAE* 4E, Doc. 13, 151–88.

Einstein, Albert, and Marcel Grossmann. 1914. "Covariance Properties of the Field Equations of the Theory of Gravitation Based on the Generalized Theory of Relativity." In *CPAE* 6E, Doc. 2, 6–15.

Einstein, Albert, and Nathan Rosen. 1937. "On Gravitational Waves." *Journal of the Franklin Institute* 223: 43–54.

Eisenstaedt, Jean. 1982. "Histoire et singularités de la solution de Schwarzschild (1915–1923)." *Archive for History of Exact Sciences* 27: 157–98.

Eisenstaedt, Jean. 1986. "La relativité générale à l'étiage: 1925–1955." *Archive for History of Exact Sciences* 35: 115–85.

Eisenstaedt, Jean. 1989a. "The Early Interpretation of the Schwarzschild Solution." In *Einstein and the History of General Relativity*, edited by D. Howard and J. Stachel. *Einstein Studies*, vol. 1 (Boston: Birkhäuser), 213–33.

Eisenstaedt, Jean. 1989b. "The Low Water Mark of General Relativity, 1925–1955." In *Einstein and the History of General Relativity*, edited by D. Howard and J. Stachel. *Einstein Studies*, vol. 1 (Boston: Birkhäuser), 277–92.

Eisenstaedt, Jean. 1993. "Lemaître and the Schwarzschild Solution." In *The Attraction of Gravitation: New Studies in the History of General Relativity*, edited by J. Earman, M. Janssen, and J. D. Norton. *Einstein Studies*, vol. 5 (Boston: Birkhauser), 353–89.

Eisenstaedt, Jean. 2006. *The Curious History of Relativity: How Einstein's Theory Was Lost and Found Again*. Princeton, NJ: Princeton University Press.

Engler, Fynn Ole, and Jürgen Renn. 2013. "Hume, Einstein und Schlick über die Objektivität der Wissenschaft." In *Moritz Schlick—Die Rostocker Jahre und ihr Einfluss auf die Wiener Zeit*, edited by F. O. Engler and M. Iven. *Schlickiana*, vol. 6 (Leipzig: Universitätsverlag), 123–56.

Friedmann, Alexander. 1922. "Über die Krümmung des Raumes." *Zeitschrift für Physik* 10, 1: 377–86.

Goenner, Hubert. 2017. "A Golden Age of General Relativity? Some Remarks on the History of General Relativity." *General Relativity and Gravitation* 49, 16pp.

Gutfreund, Hanoch, Diana K. Buchwald, and Jürgen Renn. 2016. "Gravitational Waves: Ripples in the Fabric of Spacetime Lost and Found." *Huffington Post*, February12.

Gutfreund, Hanoch , Jürgen Renn, eds. 2015a. *Relativity: The Special and the General Theory, 100th Anniversary Edition.* Princeton, NJ: Princeton University Press.

Gutfreund, Hanoch, and Jürgen Renn. 2015b. *The Road to Relativity: The History and Meaning of Einstein's "The Foundation of General Relativity" Featuring the Original Manuscript of Einstein's Masterpiece.* Princeton, NJ: Princeton University Press.

Gutfreund, Hanoch, and Jürgen Renn. 2017. *The Formative Years of Relativity: The History and Meaning of Einstein's Princeton Lectures.* Princeton, NJ: Princeton University Press.

Gutfreund, Hanoch, and Jürgen Renn. 2020. *Einstein on Einstein: Autobiographical and Scientific Reflections.* Princeton, NJ: Princeton University Press.

Hentschel, Klaus. 1997. *The Einstein Tower: An Intertexture of Dynamic Construction, Relativity Theory, and Astronomy.* Palo Alto, CA: Stanford University Press.

Hilbert, David. 1915. "Die Grundlagen der Physik (Erste Mitteilung)." *Königliche Gesellschaft der Wissenschaften zu Göttingen. Mathematisch-Physikalische Klasse. Nachrichten*, 395–407.

Hoefer, Carl. 1994. "Einstein and Mach's Principle." *Studies in History and Philosophy of Science* 25: 287–335.

Howard, Don, and John Stachel, eds. 1989. *Einstein and the History of General Relativity: Based on the Proceedings of the Osgood Hill Conference, North Andover, Massachusetts, 8–11 May 1986.* Boston: Birkhäuser.

Hubble, Edwin. 1929. "A Relation between Distance and Radial Velocity among Extra-Galactic Nebulae." *Proceedings of the National Academy of Sciences* 15(3): 168–73.

Hulse, R. A., and J. H. Taylor. 1975. "Discovery of a Pulsar in a Binary System." *Astrophysical Journal Letters* 195, L51–L53.

Infeld, Leopold, and Jerzy Plebanski. 1960. *Motion and Relativity.* Oxford: Pergamon.

Janssen, Michel. 2005. "Of Pots and Holes: Einstein's Bumpy Road to General Relativity." *Annalen der Physik* 14: Supplement 58–85. Reprinted in *Einstein's Annalen Papers. The Complete Collection 1901–1922*, edited by J. Renn (Weinheim: Wiley-VCH).

Janssen, Michel. 2014a. "Appendix: Special Relativity." In *The Cambridge Companion to Einstein*, edited M. Janssen and C. Lehner (New York: Cambridge University Press), 455–506.

Janssen, Michel. 2014b. "No Success Like Failure . . .': Einstein's Quest for General Relativity, 1907–1920." In *The Cambridge Companion to Einstein*, edited by M. Janssen and C. Lehner (New York: Cambridge University Press), 167–227.

Janssen, Michel, and Christoph Lehner, eds. 2014. *The Cambridge Companion to Einstein.* Cambridge: Cambridge University Press.

Janssen, Michel and Jürgen Renn. 2007. "Untying the Knot: How Einstein Found His Way Back to Field Equations Discarded in the Zurich Notebook." In *Einstein's Zurich Notebook: Commentary and Essays*, edited by M. Janssen, J. Norton, J. Renn, T. Sauer, and J. Stachel. *The Genesis of General Relativity*, vol. 2 (Dordrecht: Springer), 839–925.

Janssen, Michel, and Jürgen Renn. 2015a. "Arch and Scaffold: How Einstein Found His Field Equations." *Physics Today* 68, 11: 30–36.

Janssen, Michel, and Jürgen Renn. 2015b. "Einstein Was No Lone Genius." *Nature* 572: 298–301.

Janssen, Michel, and Jürgen Renn. Forthcoming. *How Einstein Found His Field Equations. A Source Book.* Heidelberg: Springer-Verlag.

Kaiser, David. 2005. *Drawing Theories Apart: The Dispersion of Feynman Diagrams in Postwar Physics.* Chicago: University of Chicago Press.

Kaiser, David. 2012. "Booms, Busts, and the World of Ideas: Enrollment Pressures and the Challenge of Specialization." *Osiris* 27: 276–302.

Kennefick, Daniel. 2005. Einstein Versus the *Physical Review:* A Great Scientist Can Benefit from Peer Review, Even While Refusing to Have Anything to Do with It." *Physics Today* 58, 43–48.

Kennefick, Daniel. 2007. *Traveling at the Speed of Thought: Einstein and the Quest for Gravitational Waves.* Princeton, NJ: Princeton University Press.

Kennefick, Daniel. 2012. "Not Only Because of Theory: Dyson, Eddington, and the Competing Myths of the 1919 Eclipse Expedition." In *Einstein and the Changing Worldviews of Physics,* edited by C. Lehner, J. Renn, and M. Schemmel. *Einstein Studies,* vol. 12. (Boston: Birkhäuser), 201–32.

Kennefick, Daniel. 2017. "The Binary Pulsar and the Quadrupole Formula Controversy." *European Physical Journal H* 42(2), 293–310.

Kragh, Helge S. 1996. *Cosmology and Controversy: The Historical Development of Two Theories of the Universe.* Princeton, NJ: Princeton University Press.

Kramer, David. 2016. "Gravitational Waves Are Detected for First Time." *Physics Today.* doi:10.1063/PT.5.9053.

Kretschmann, Erich. 1917. "Über den physikalischen Sinn der Relativitätspostulate, A. Einsteins neue und seine ursprüngliche Relativitätstheorie." *Annalen der Physik* 53(16), 575–614.

Lalli, Roberto. 2016. " 'Dirty Work,' but Someone Has to Do It: Howard P. Robertson and the Refereeing Practices of *Physical Review* in the 1930s." *Records and Notes* 70: 151–74.

Lalli, Roberto. 2017. *Building the General Relativity and Gravitation Community During the Cold War. Springer Briefs in History of Science and Technology.* Cham: Springer International.

Lemaître, Georges. 1927. "Un Univers homogène de masse constante et de rayon croissant rendant compte de la vitesse radiale des nébuleuses extra-galactiques." *Annales de la Société Scientifique de Bruxelles* 47, 49.

Mach, Ernst. 1989 [1883]. *The Science of Mechanics: A Critical and Historical Account of Its Development.* LaSalle, IL: Open Court.

Mercier, André, and Michel A. Kervaire. 1956. *Fünfzig Jahre Relativitätstheorie, Verhandlungen—Cinquantenaire de la théorie de la relativité, Actes— Jubilee of Relativity Theory, Proceedings.* Boston: Birkhäuser.

Mie, Gustav. 1913. "Grundlagen einer Theorie der Materie." *Annalen der Physik* 37, 511–34. English translation: "Foundations of a Theory of Matter (Excerpts)," in *From an Electromagnetic Theory of Matter to a New Theory of Gravitation,* edited by J. Renn and M. Schemmel. *The Genesis of General Relativity,* vol. 4 (Dordrecht: Springer, 2007), 634–97.

Norton, John. 1984. "How Einstein Found His Field Equations, 1912–1915." *Historical Studies in the Physical Sciences* 14, 253–316. Reprinted in Howard and Stachel 1989, 101–59.

Norton, John. 2014. "Einstein's Special Theory of Relativity and the Problems in the Electrodynamics of Moving Bodies That Led Him to It." In *The Cambridge Companion to Einstein,* edited by M. Janssen and C. Lehner (New York: Cambridge University Press), 72–102.

Renn, Jürgen. 2004. "The Relativity Revolution from the Perspective of Historical Epistemology." *Isis* Special Issue: The Elusive Icon: Albert Einstein 1905–2005, 640–48.

Renn, Jürgen. 2005. "Before the Riemann Tensor: The Emergence of Einstein's Double Strategy." In *The Universe of General Relativity,* edited by J. Eisenstaedt and A. J. Kox. *Einstein Studies,* vol. 11 (Boston: Birkhäuser), 53–65.

Renn, Jürgen. 2006. *Auf den Schultern von Riesen und Zwergen: Einsteins unvollendete Revolution.* Weinheim: Wiley-VCH.

Renn, Jürgen. 2007a. "Classical Physics in Disarray: The Emergence of the Riddle of Gravitation." In *Einstein's Zurich Notebook: Introduction and Source,* edited by M. Janssen et al. *The Genesis of General Relativity,* vol. 1 (Dordrecht: Springer), 21–80.

Renn, Jürgen. ed. 2007b. *The Genesis of General Relativity* (4 vols.). Boston Studies in the Philosophy of Science, vol. 250. Dordrecht: Springer.

Renn, Jürgen. 2007c. "The Summit Almost Scaled: Max Abraham as Pioneer of a Relativistic Theory of Gravitation." In *Gravitation in the Twilight of Classical Physics: Between Mechanics, Field Theory, and Astronomy,* edited by J. Renn and M. Schemmel. *The Genesis of General Relativity,* vol. 3 (Dordrecht: Springer), 305–30.

Renn, Jürgen. 2007d. "The Third Way to General Relativity." in *Gravitation in the Twilight of Classical Physics: Between Mechanics, Field Theory, and Astronomy,* edited by J. Renn and M. Schemmel. *The Genesis of General Relativity,* vol. 3 (Dordrecht: Springer), 21–75.

Renn, Jürgen. Forthcoming. *On the Shoulders of Giants and Dwarfs: Einstein's Unfinished Revolution.* Princeton, NJ: Princeton University Press.

Renn, Jürgen, and Robert Rynasiewicz. 2014. "Einstein's Copernican Revolution." In *The Cambridge Companion to Einstein,* edited by M. Janssen and C. Lehner (New York: Cambridge University Press), 38–71.

Renn, Jürgen, and Tilman Sauer. 1999. "Heuristics and Mathematical Representation in Einstein's Search for a Gravitational Field Equation." In *The Expanding Worlds of General Relativity,* edited by H. Goenner et al., *Einstein Studies,* vol. 7 (Boston: Birkhäuser), 87–125.

Renn, Jürgen, and Tilman Sauer. 2003. "Eclipses of the Stars—Mandl, Einstein, and the Early History of Gravitational Lensing." In *Revisiting the Foundations of Relativistic Physics—Festschrift in Honour of John Stachel,* edited by A. Ashtekar et al. (Dordrecht: Kluwer Academic), 69–92.

Renn, Jürgen, and Tilman Sauer. 2007. "Pathways Out of Classical Physics: Einstein's Double Strategy in Searching for the Gravitational Field Equation." In *Einstein's Zurich Notebook: Introduction and Source,* edited by M. Janssen et al. *The Genesis of General Relativity,* vol. 1 (Dordrecht: Springer), 113–312.

Renn, Jürgen, Tilman Sauer, and John Stachel. 1997. "The Origin of Gravitational Lensing. A Postscript to Einstein's 1936 Science Paper." *Science* 275, 5297: 184–86.

Renn, Jürgen, and Matthias Schemmel. 2012. "Theories of Gravitation in the Twilight of Classical Physics." In *Einstein and the Changing Worldviews of Physics,* edited by C. Lehner, J. Renn, and M. Schemmel. *Einstein Studies,* vol. 12 (New York: Springer), 3–22.

Renn, Jürgen, and John Stachel. 2007. "Hilbert's Foundation of Physics: From a Theory of Everything to a Constituent of General Relativity." In *Gravitation in the Twilight of Classical Physics: The Promise of Mathematics,* edited by J. Renn and M. Schemmel. *The Genesis of General Relativity,* vol. 4 (Dordrecht: Springer), 857–973.

Rosen, Nathan. 1937. "Plane Polarized Waves in the General Theory of Relativity." *Physikalische Zeitschrift der Sowjetunion* 12: 366–72.

Rosen, Nathan. 1956. "Gravitational Waves." In *Fünfzig Jahre Relativitätstheorie, Verhandlungen—Cinquantenaire de la théorie de la relativité, Actes—Jubilee of Relativity Theory, Proceedings,* edited by A. Mercier and M. A. Kervair (Boston: Birkhäuser), 171–75.

Rynasiewicz, Robert. 1999. "Kretschmann's Analysis of Covariance and Relativity Principles." In *The Expanding Worlds of General Relativity,* edited by H. Goenner et al., *Einstein Studies,* vol. 7. (Boston: Birkhäuser), 431–62.

Sauer, Tilman. 2005a. "Einstein Equations and Hilbert Action: What Is Missing on Page 8 of the Proofs for Hilbert's First Communication on the Foundations of Physics?" *Archive for History of Exact Sciences* 59, 577–90.

Sauer, Tilman. 2005b. "Einstein's Review Paper on General Relativity Theory." In *Landmark Writings in Western Mathematics, 1640–1940,* edited by I. Grattan-Guinness (Amsterdam: Elsevier), 802–22.

Sauer, Tilman. 2008. "Nova Geminorum 1912 and the Origin of the Idea of Gravitational Lensing." *Archive for History of Exact Sciences* 62, 1–22.

Sauer, Tilman. 2014. "Marcel Grossmann and His Contribution to the General Theory of Relativity." arXiv:1312.4068. Published 2015 in *Proceedings of the 13th Marcel Grossmann Meeting on Recent Developments in Theoretical and Experimental General Relativity, Gravitation, and Relativistic Field Theory,* edited by K. Rosquist, R. T. Jantzen, and R. Ruffinn (Singapore: World Scientific), 456–503.

Schemmel, Matthias. 2005. "An Astronomical Road to General Relativity: The Continuity between Classical and Relativistic Cosmology in the Work of Karl Schwarzschild." *Science in Context* 18, 451–78. Reprinted in Renn (2007b, vol. 3).

Schutz, Bernard. F. 2012. "Thoughts About a Conceptual Framework for Relativistic Gravity." In *Einstein and the Changing Worldviews of Physics,* edited by C. Lehner, J. Renn, M. Schemmel, *Einstein Studies* vol. 12 (Boston: Birkhäuser), 259–69.

Stachel, John. 1989a. "Einstein's Search for General Covariance, 1912–1915." In *Einstein and the History of General Relativity,* edited by D. Howard and J. Stachel, *Einstein Studies* vol. 1 (Boston: Birkhäuser), 63–100.

Stachel, John. 1989b. "The Rigidly Rotating Disk as the 'Missing Link' in the History of General Relativity." In *Einstein and the History of General Relativity*, edited by D. Howard and J. Stachel. *Einstein Studies* vol. 1 (Boston: Birkhäuser), 48–62.

Stachel, John. 2007. "The First Two Acts. Prologue: The Development of General Relativity, a Drama in Three Acts." In *Einstein's Zurich Notebook: Introduction and Source*, edited by M. Janssen et al. *The Genesis of General Relativity*, vol. 1 (Dordrecht: Springer), 81–111.

Stachel, John. 2014. "The Hole Argument and Some Physical and Philosophical Implications." *Living Reviews in Relativity* 17, 1. https://doi.org/10.12942/lrr-2014-1.

Taylor, J. H., and J. M. Weisberg. 1982. "A New Test of General Relativity—Gravitational Radiation and the Binary Pulsar PSR 1913+16." *Astrophysical Journal* 253, 908–920.

Thorne, Kip S. 1994. *Black Holes and Time Warps: Einstein's Outrageous Legacy*. New York: Norton.

Toennies, J. Peter, Horst Schmidt-Bocking, Bretislav Friedrich, and Julian C. A. Lower. 2011. "Otto Stern (1888–1969): The Founding Father of Experimental Atomic Physics," *Annals of Physics (Berlin)* 523(12), 1045–70.

Trimble, Virginia. 2017. "Wired by Weber: The Story of the First Searcher and Searches for Gravitational Waves." *European Physical Journal H* 42(2): 261–91.

van Dongen, Jeroen. 2010. *Einstein's Unification*. Cambridge: Cambridge University Press.

Weber, J. 1969. "Evidence for Discovery of Gravitational Radiation." *Physical Review Letters* 22, 1320–24.

Wheeler, John Archibald. 1998. *Geons, Black Holes, and Quantum Foam: A Life in Physics*. New York: Norton.

Will, Clifford. M. 1986. "*Was Einstein Right? Putting General Relativity to the Test*. New York: Basic Books.

Will, Clifford. M. 1989. "The Renaissance of General Relativity" In *The New Physics*, edited by P. Davies (Cambridge: Cambridge University Press), 7–33.

Collins: The Detection of Gravitational Waves

Bartusiak, Marcia. 2000. *Einstein's Unfinished Symphony*. Washington, DC: Joseph Henry.

Collins, Harry. 2004. *Gravity's Shadow: The Search for Gravitational Waves*. Chicago: University of Chicago Press.

Collins, Harry. 2017. *Gravity's Kiss: The Detection of Gravitational Waves*. Cambridge, MA: MIT Press.

Kennefick, Daniel. 2007. *Traveling at the Speed of Thought: Einstein and the Quest for Gravitational Waves*, Princeton, NJ: Princeton University Press.

Taylor, J. H., and J. M. Weisberg. 1982. "A New Test of General Relativity—Gravitational Radiation and the Binary Pulsar PSR 1913+16." *Astrophysical Journal* 253: 908–920; doi:10.1086/159690.

Weisberg, J. M., and J. H. Taylor. 2004. "Relativistic Binary Pulsar B1913+16: Thirty years of Observations and Analysis." https://arxiv.org/abs/astro-ph/0407149.

Winch, Peter G. 1958. *The Idea of a Social Science*. London: Routledge and Kegan Paul.

Wittgenstein, Ludwig. 1953. *Philosophical Investigations*. Oxford: Blackwell.

Howard: How General Relativity Shaped Twentieth-Century Philosophy of Science

Avenarius, Richard. 1888–1890. *Kritik der reinen Erfahrung*. 2 vols. Leipzig: Fues (R. Reisland).

Beller, Mara. 2000. "Kant's Impact on Einstein's Thought." In *Einstein: The Formative Years*, edited by Don Howard and John Stachel. Boston: Birkhäuser, 83–106.

Cahan, David. 1994. "Anti-Helmholtz, Anti-Zöllner, Anti-Dühring: The Freedom of Science in Germany during the 1870s." In *Universalgenie Helmholtz. Rückblick nach 100 Jahren*, edited by Lorenz Krüger. Berlin: Akademie Verlag, 330–44.

Canales, Jimena. 2015. *The Physicist and the Philosopher: Einstein, Bergson, and the Debate That Changed Our Understanding of Time*. Princeton, NJ: Princeton University Press.

Carnap, Rudolf. 1921. *Der Raum. Ein Beitrag zur Wissenschaftslehre*. Inaugural-Dissertation zur Erlangung der Doktorwürde der hohen philosophischen Fakultät der Universität Jena. Göttingen: Dieterich'schen Univ.-Buchdruckerei, W. Fr. Kaestner. Reprinted as "Kant-Studien" Ergänzungshefte, no. 56. Berlin: Reuther & Reichard, 1922.

Carnap, Rudolf. 1923. "Über die Aufgabe der Physik." *Kant-Studien* 28, 90–107.

Carnap, Rudolf. 1925. "Über die Abhängigkeit der Eigenschaften des Raumes von denen der Zeit." *Kant-Studien* 30, 331–45.

Carnap, Rudolf. 1926. *Physikalische Begriffsbildung*. Wissen und Wirken. Einzelschriften zu den Grundfragen des Erkennens und Schaffens. edited by Emil Ungerer, vol. 39. Karlsruhe: G. Braun.

Carnap, Rudolf. 1927. "Eigentliche und Uneigentliche Begriffe." *Symposion. Philosophische Zeitschrift für Forschung und Ausprache* 1, 355–374.

Carnap, Rudolf. 1928. *Der logische Aufbau der Welt*. Berlin: Weltkreis.

Carnap, Rudolf. 1929. *Abriss der Logistik. Mit besonderer Berücksichtigung der Relationstheorie und ihrer Anwendungen*. Schriften zur Wissenschaftlichen Weltauffassung, vol. 2. edited by Philipp Frank and Moritz Schlick. Vienna: Julius Springer.

Carnap, Rudolf. 1963. "Intellectual Autobiography." In *The Philosophy of Rudolf Carnap*, edited by Paul Arthur Schilpp. The Library of Living Philosophers, vol. 11. La Salle, IL: Open Court, 1–84.

Carnap, Rudolf. 1977. *Two Essays on Entropy*, edited by Abner Shimony. Berkeley: University of California Press.

Cassirer, Ernst. 1910. *Substanzbegriff und Funktionsbegriff. Untersuchungen über die Grundfragen der Erkenntniskritik*. Berlin: Bruno Cassirer.

Cassirer, Ernst. 1921. *Zur Einsteinschen Relativitätstheorie. Erkenntnistheoretische Betracht-ungen*. Berlin: Bruno Cassirer.

Cassirer, Ernst. 1937. *Determinismus und Indeterminismus in der modernen Physik. Historische und systematische Studien zum Kausalproblem*. Göteborg: Elanders Boktryckeri Aktiebolag.

Cohen, Hermann. 2015. *Briefe an August Stadler*, edited by Hartwig Wiedebach. Basel: Schwabe.

Dedekind, Richard. 1888. *Was sind und was sollen die Zahlen?* Braunschweig: Friedrich Vieweg. 2nd ed. 1893.

Dubislav, Walter. 1929. "Joseph Petzoldt in memoriam: Vortrag, gehalten am 15. Oktober 1929 in der Gesellschaft für empirische Philosophie, Ortsgruppe Berlin." *Annalen der Philosophie und der philosophischen Kritik* 8, 289–95.

Duhem, Pierre. 1906. *La Théorie physique: son objet et sa structure*. Paris: Chevalier & Rivière. Page numbers from the English translation of the second edition (Duhem 1914): *The Aim and Structure of Physical Theory*. Philip P. Wiener, trans. Princeton: Princeton University Press, 1954.

Duhem, Pierre. 1908. *Ziel und Struktur der physikalischen Theorien*. Friedrich Adler, trans. Foreword by Ernst Mach. Leipzig: Johann Ambrosius Barth.

Duhem, Pierre. 1914. *La Théorie physique: son objet et sa structure*, 2nd ed. Paris: Marcel Rivière & Cie.

Dühring, Eugen. 1887. *Kritische Geschichte der allgemeinen Principien der Mechanik*, 3rd ed. Leipzig: Fues's Verlag (R. Reisland).

Einstein, Albert. 1914. "Die formale Grundlage der allgemeinen Relativitätstheorie." *Königlich Preussisch Akademie der Wissenschaften* (Berlin). *Sitzungsberichte*, 1030–1085.

Einstein, Albert. 1915a. "Zur allgemeinen Relativitätstheorie." *Königlich Preussische Akademie der Wissenschaften* (Berlin). *Sitzungsberichte*, 778–86, 799–801.

Einstein, Albert. (1915b. "Erklärung der Perihelbewegung des Merkur aus der allgemeinen Relativitäts- theorie." *Königlich Preussische Akademie der Wissenschaften* (Berlin). *Sitzungsberichte*, 831–39.

Einstein, Albert. 1915c. "Die Feldgleichungen der Gravitation." *Königlich Preussische Akademie der Wissenschaften* (Berlin). *Sitzungsberichte*, 844–47.

Einstein, Albert. 1916a. "Die Grundlage der allgemeinen Relativitätstheorie." *Annalen der Physik* 49, 769–822.

Einstein, Albert. 1916b. "Ernst Mach." *Physikalische Zeitschrift* 17(1916), 101–4.

Einstein, Albert. 1921. *Geometrie und Erfahrung. Erweiterte Fassung des Festvortrages gehalten an der Preussischen Akademie der Wissenschaften zu Berlin am 27. Januar 1921.* Berlin: Julius Springer. Page numbers and quotations from the translation in Einstein 1954, 232–46.

Einstein, Albert. 1924. Review of Elsbach 1924. *Deutsche Literaturzeitung* 45, 1685–92.

Einstein, Albert. 1928. "A propos de 'la déduction relativiste' de M. Émile Meyerson." André Metz, trans. *Revue philosophique de la France et de l'étranger* 105, 161–66.

Einstein, Albert. 1936. "Physik und Realität." *Journal of The Franklin Institute* 221, 313–47.

Einstein, Albert. 1949. "Remarks Concerning the Essays Brought Together in This Co-operative Volume." In Schilpp 1949, 665–88.

Einstein, Albert. 1954. *Ideas and Opinions.* edited by Carl Seelig, translated by Sonja Bargmann. New York: Bonanza Books.

Einstein, Albert and Marcel Grossmann. 1913. *Entwurf einer verallgemeinerten Relativitätstheorie und einer Theorie der Gravitation. I. Physikalischer Teil von Albert Einstein. II. Mathematischer Teil von Marcel Grossmann.* Leipzig and Berlin: B.G. Teubner. Reprinted with added "Bemerkungen," *Zeitschrift für Mathematik und Physik* 62(1914), 225–61.

Einstein, Albert, and Marcel Grossmann. 1914. "Kovarianzeigenschaften der Feldgleichungen der auf die verallgemeinerte Relativitätstheorie gegründeten Gravitationstheorie." *Zeitschrift für Mathematik und Physik* 63, 215–225.

Elsbach, Alfred. 1924. *Kant und Einstein. Untersuchungen über das Verhältnis der modernen Erkenntnistheorie zur Relativitätstheorie.* Berlin and Leipzig: Walter de Gruyter.

Frank, Philipp. 1917. "Die Bedeutung der physikalischen Erkenntnistheorie Machs für das Geistesleben der Gegenwart." *Die Naturwissenschaften* 5, 65–72.

Frank, Philipp. 1932. *Das Kausalgesetz und seine Grenzen.* Vienna: Julius Springer.

Frank, Philipp. 1938. *Interpretations and Misinterpretations of Modern Physics.* Paris: Hermann & cie.

Frank, Philipp. 1949a. *Einstein: His Life and Times.* New York: Alfred A. Knopf.

Frank, Philipp. 1949b. "Einstein's Philosophy of Science." *Reviews of Modern Physics* 21, 349–55.

Friedman, Michael. 2001. *Dynamics of Reason: The 1999 Kant Lectures at Stanford University.* Stanford, CA: CSLI Publications.

Gawronsky, Dimitry. 1949. "Ernst Cassirer: His Life and His Work." In *The Philosophy of Ernst Cassirer,* edited by Paul Arthur Schilpp. Evanston, IL: The Library of Living Philosophers, 3–37.

Giovanelli, Marco. 2003. *August Stadler: Interprete di Kant.* Naples: Guida.

Hecht, Hartmut, and Dieter Hoffmann. 1982. "Die Berufung Hans Reichenbachs an die Berliner Universität." *Deutsche Zeitschrift für Philosophie* 30, 651–62.

Hilbert, David. 1899. *Grundlagen der Geometrie.* Leipzig: B.G. Teubner.

Howard, Don. 1984. "Realism and Conventionalism in Einstein's Philosophy of Science: The Einstein-Schlick Correspondence." *Philosophia Naturalis* 21, 618–29.

Howard, Don. 1990. "Einstein and Duhem." In *Pierre Duhem: Historian and Philosopher of Science.* Proceedings of the Conference at Virginia Polytechnic Institute and State University, Blacksburg, Virginia, March 16–18, 1989. Roger Ariew and Peter Barker, eds. *Synthese* 83, 363–84.

Howard, Don. 1991. "Einstein and *Eindeutigkeit*: A Neglected Theme in the Philosophical Background to General Relativity." In *Historical Studies in General Relativity,* edited by Jean Eisenstaedt and A. J. Kox. Boston: Birkhäuser, 154–243.

Howard, Don. 1993. "Was Einstein Really a Realist?" *Perspectives on Science: Historical, Philosophical, Social* 1, 204–51.

Howard, Don. 1994. "Einstein, Kant, and the Origins of Logical Empiricism." In *Language, Logic, and the Structure of Scientific Theories,* edited by Wesley Salmon and Gereon Wolters. Pittsburgh: University of Pittsburgh Press; Konstanz: Universitätsverlag, 45–105.

Howard, Don. 1996. "Relativity, *Eindeutigkeit,* and Monomorphism: Rudolf Carnap and the Development of the Categoricity Concept in Formal Semantics." In *Origins of Logical Empiricism,* edited by Ronald N. Giere and Alan Richardson. Minneapolis and London: University of Minnesota Press, 1996, 115–64.

Howard, Don. 1999. "Point Coincidences and Pointer Coincidences: Einstein on Invariant Structure in Spacetime Theories." In *History of General Relativity IV: The Expanding Worlds of General Relativity,* edited by Hubert Goenner et al. (Boston: Birkhäuser, 1999), 463–500.

Howard, Don. 2004. "Fisica e filosofia della scienza all'alba del XX secolo." In *Storia della scienza.* Vol. 8, *La Seconda revoluzione scientifica,* edited by Umberto Bottazzini et al. Rome: Istituto della Enciclopedia Italiana, 3–16.

Howard, Don. 2005. "Albert Einstein as a Philosopher of Science." *Physics Today* 58, no. 11, 34–40.

Howard, Don. 2010. "'Let me briefly indicate why I do not find this standpoint natural.' Einstein, General Relativity, and the Contingent A Priori." In *Discourse on a New Method: Reinvigorating the Marriage of History and Philosophy of Science,* edited by Michael Dickson and Mary Domski. Chicago: Open Court, 333–55.

Howard, Don. 2014. "Einstein and the Development of Twentieth-Century Philosophy of Science." In *The Cambridge Companion to Einstein,* edited by Michel Janssen and Christoph Lehner. New York: Cambridge University Press, 354–76.

Howard, Don. Forthcoming. "The Philosopher-Physicists: Albert Einstein and Philipp Frank. In *Logical Empiricism and the Physical Sciences from Philosophy of Nature to Philosophy of Physics,* edited by Sebastian Lutz and Adam Tamas Tuboly. Cham, Switzerland: Springer.

Howard, Don, and John Norton. 1994. "Out of the Labyrinth? Einstein, Hertz, and the Göttingen Answer to the Hole Argument." In *The Attraction of Gravitation: New Studies in the History of General Relativity,* edited by John Earman, Michel Janssen, and John Norton. Boston: Birkhäuser, 1994, 30–62.

Howard, Don, and John Stachel, eds. 1989. *Einstein and the History of General Relativity: Based on the Proceedings of the Osgood Hill Conference, North Andover, Massachusetts, 8–11 May 1986.* Boston: Birkhäuser.

Hume, David. 1739. *A Treatise of Human Nature.* London: John Noon.

Hume, David. 1895. *Ein Traktat über die menschliche Natur: Ein Versuch die Methode der Erfahrung in die Geisteswissenschaften einzuführen.* Theodore Lipps, trans. Vol. 1. *Üben den Verstand.* Hamburg: Voss; 2nd enl. ed., 1904.

Janssen, Michel, and Jürgen Renn. 2015. "Arch and Scaffold: How Einstein Found His Field Equations." *Physics Today* 68 (November), 30–36.

Kant, Immanuel. 1878a. *Kritik der reinen Vernunft. Text der Ausgabe 1781 mit Beifügung sämmtlicher Abweichungen der Ausgabe 1787,* 2nd ed. Edited by Karl Kehrbach. Leipzig: Philipp Reclam jun.

Kant, Immanuel. 1878b. *Kritik der praktischen Vernunft. Text der Ausgabe 1788, (A) unter Berücksichtigung der 2. Ausgabe 1792 (B) und der 4. Ausgabe 1797 (D).* Edited by Karl Kehrbach. Leipzig: Philipp Reclam jun.

Kant, Immanuel. 1878c. *Kritik der Urtheilskraft. Text der Ausgabe 1790, (A) mit Beifügung sämmtlicher Abweichungen der Ausgaben 1793 (B) und 1799 (C).* Edited by Karl Kehrbach. Leipzig: Philipp Reclam jun.

Knott, Robert. 1907. "Rosenberger, Ferdinand." In *Allgemeine Deutsche Biographie.* Vol. 53. Leipzig: Duncker & Humblot, 495–96.

Köhnke, Klaus Christian. 1986. *Entstehung und Aufstieg des Neukantianismus. Die deutsche Universitätsphilosophie zwischen Idealismus und Positivismus.* Frankfurt am Main: Suhrkamp.

Lange, Friedrich Albert. 1866. *Geschichte des Materialismus und Kritik seiner Bedeutung in der Gegenwart.* 2 vols. Iserlohn: J. Baedeke. 2nd ed. 1873–1875.

Lange, Friedrich Albert. 1905. *Geschichte des Materialismus und Kritik seiner Bedeutung in der Gegenwart.* Iserlohn: Baedeker. [Reprint of the 2nd. ed.]

Lange, Friedrich Albert. 1905. *Geschichte des Materialismus und Kritik seiner Bedeutung in der Gegenwart.* Book 1, *Geschichte des Materialismus bis auf Kant.* Edited by O. A. Ellissen. Leipzig: Philipp Reclam jun.

Le Roy, Èdouard. 1901. "Un positivisme nouveau." *Revue de Métaphysique et de Morale* 9, 138–53.

Liebmann, Otto. 1865. *Kant und die Epigonen. Eine kritische Abhandlung.* Stuttgart: Carl Schober. 2nd ed. Edited by Bruno Bauch. Berlin: Reuther & Reichard, 1912.

Mach, Ernst. 1896. *Die Principien der Wärmelehre. Historisch-kritisch entwickelt.* Leipzig: Johann Ambrosius Barth.

Mach, Ernst. 1897. *Die Mechanik in ihrer Entwickelung historisch-kritisch dargestellt*, 3rd impr. and enl. ed. Leipzig: Brockhaus.

Mach, Ernst. 1900. *Die Analyse der Empfindungen und das Verhältnis des Physischen zum Psychischen*. 2nd. ed. Jena: G. Fischer; 3rd enl. ed. 1902; 4th enl. ed. 1903.

Meyerson, Émile. 1925. *La déduction relativiste*. Paris: Payot. English trans.: *The Relativistic Deduction: Epistemological Implications of the Theory of Relativity*. Milič Čapek, trans. Dordrecht and Boston: D. Reidel, 1985.

Mill, John Stuart. 1872. *A System of Logic, Ratiocinative and Inductive: Being a Connected View of the Principles of Evidence and the Methods of Scientific Investigation*. London: Longmans, Green, Reader, and Dyer.

Mill, John Stuart. 1877. *System der deductiven und inductiven Logik. Eine Darlegung der principien wissenschaftlicher Forschung, insbesondere der Naturforschung*. J. Schiele, trans. Braunschweig: Friedrich Vieweg und Sohns.

Mill, John Stuart. 1884–1886. *System der deductiven und inductiven Logik. Eine Darlegung der Grundsätze der Beweislehre und der Methoden wissenschaftlicher Forschung*. Theodore Gomperz, trans. Leipzig: Fues.

Norton, John. 1984. "How Einstein Found His Field Equations, 1912–1915." *Historical Studies in the Physical Sciences* 14, 253–316. Reprinted in Howard and Stachel 1989, 101–59.

Pearson, Karl. 1892. *The Grammar of Science*. London: Walter Scott; New York: Charles Scribner's Sons. 2nd ed. London: Adam and Charles Black, 1900.

Petzoldt, Joseph. 1895. "Das Gesetz der Eindeutigkeit." *Vierteljahrsschrift für wissenschaftliche Philosophie und Soziologie* 19, 146–203.

Petzoldt, Joseph. 1906. *Das Weltproblem von positivistischem Standpunkte aus*. Aus Natur und Geisteswelt, no. 133. Leipzig: B. G. Teubner.

Petzoldt, Joseph. 1912a. "Die Relativitätstheorie im erkenntnistheoretischer Zusammenhange des relativistischen Positivismus." *Deutsche Physikalische Gesellschaft. Verhandlungen* 14, 1055–64.

Petzoldt, Joseph. 1912b. *Das Weltproblem vom Standpunkte des relativistischen Positivismus aus, historisch-kritisch dargestellt*, 2nd ed. Wissenschaft und Hypothese, vol. 14. Leipzig and Berlin: B. G. Teubner. [2nd ed. of Petzoldt 1906.]

Petzoldt, Joseph. 1914. "Die Relativitätstheorie der Physik." *Zeitschrift für positivistische Philosophie* 2, 1–53.

Poincaré, Henri. 1902a. "Sur la valeur objective des théories physiques." *Revue de Métaphysique et de morale* 10(1902), 263–93.

Poincaré, Henri. 1902b. *La science et l'hypothèse*. Paris: Ernest Flammarion.

Poincaré, Henri. 1904. *Wissenschaft und Hypothese*. Ferdinand Lindemann, trans. Leipzig: B.G. Teubner.

Quine, Willard Van Orman. 1951. "Two Dogmas of Empiricism." *Philosophical Review* 60, 29–43. Reprinted in *From a Logical Point of View: 9 Logico-Philosophical Essays*. Cambridge, MA: Harvard University Press, 1953, 20–46.

Quine, Willard Van Orman. 1960. *Word and Object*. Cambridge, MA: MIT Press.

Reichenbach, Hans. 1916. *Der Begriff der Wahrscheinlichkeit für die mathematische Darstellung der Wirklichkeit*. Leipzig: Johann Ambrosius Barth.

Reichenbach, Hans. 1920. *Relativitätstheorie und Erkenntnis Apriori*. Berlin: Julius Springer.

Reichenbach, Hans. 1922. "Die gegenwärtige Stand der Relativitätsdiskussion." *Logos* 10: 316–78. Page numbers and quotations from the translation in Reichenbach 1978, vol. 2, 3–47.

Reichenbach, Hans. 1924. *Axiomatik der relativistischen Raum-Zeit-Lehre*. Braunschweig: Friedrich Vieweg und Sohn. Page numbers and quotations from the English translation: *Axiomatization of the Theory of Relativity*, edited and translated by Maria Reichenbach. Berkeley and Los Angeles: University of California Press, 1969.

Reichenbach, Hans. 1928. *Philosophie der Raum-Zeit-Lehre*. Berlin and Leipzig: Walter de Gruyter. Page numbers and quotations from the English translation: *Philosophy of Space & Time*. Maria Reichenbach and John Freund, trans. New York: Dover, 1957.

Reichenbach, Hans. 1944. *Philosophic Foundations of Quantum Mechanics*. Berkeley and Los Angeles: University of California Press.

Reichenbach, Hans. 1949. "The Philosophical Significance of the Theory of Relativity." In Schilpp 1949, 289–311.

Reichenbach, Hans. 1978. *Selected Writings, 1909–1953,* 2 vols. edited by Maria Reichenbach and Robert S. Cohen. Translated by Elizabeth Hughes Schneewind. Vienna Circle Collection, vol. 4. Dodrecht and Boston: D. Reidel.

Renn, Jürgen, et al. 2007. *The Genesis of General Relativity.* 4 Vols. Dordrecht: Springer.

Richardson, Alan W. 1992. "Idealism and Carnap's Construction of the World." *Synthese* 93, 59–92.

Rosenberger, Ferdinand. 1895. *Isaac Newton und seine physikalischen Prinzipien. Ein Hauptstück aus der Entwickelungsgeschichte der modernen Physik.* Leipzig: Johann Ambrosius Barth.

Rosenthal-Schneider, Ilse. 1980. *Reality and Scientific Truth: Discussions with Einstein, von Laue, and Planck,* edited by Thomas Braun. Detroit, MI: Wayne State University Press.

Ryckman, Thomas. 2005. *The Reign of Relativity: Philosophy in Physics 1915–1925.* New York: Oxford University Press.

Schilpp, Paul Arthur, ed. 1949. *Albert Einstein: Philosopher-Scientist.* The Library of Living Philosophers, vol. 7. Evanston, IL: The Library of Living Philosophers.

Schlick, Moritz. 1904. *Über die Reflexion des Lichtes in einer inhomogenen Schicht.* Berlin: Universitäts-buchdruckerei von G. Schade (O. Francke).

Schlick, Moritz. 1910. "Das Wesen der Wahrheit nach der modernen Logik." *Vierteljahrsschrift für wissenschaftliche Philosophie und Soziologie* 34, 386–477.

Schlick, Moritz. 1915. "Die philosophische Bedeutung des Relativitätsprinzips." *Zeitschrift für Philosophie und philosophische Kritik* 159, 129–75.

Schlick, Moritz. 1917. *Raum und Zeit in der gegenwärtigen Physik. Zur Einführung in das Verständnis der allgemeinen Relativitätstheorie.* Berlin: Julius Springer.

Schlick, Moritz. 1918. *Allgemeine Erkenntnislehre.* Berlin: Julius Springer.

Schlick, Moritz. 1920. *Space and Time in Contemporary Physics: An Introduction to the Theory of Relativity and Gravitation.* Henry Brose, trans. Oxford: Clarendon.

Schlick, Moritz. 1921. "Kritizistische oder empiristische Deutung der neuen Physik." *Kant-Studien* 26, 96–111. Page numbers and quotations from the translation in Schlick 1979, vol. 1, 322–34.

Schlick, Moritz. 1922. "Review of Reichenbach 1920." *Die Naturwissenschaften* 10, 873–74.

Schlick, Moritz. 1925. *Allgemeine Erkenntnislehre,* 2nd ed. Berlin: Julius Springer. Reprint Frankfurt am Main: Suhrkamp, 1979. Page numbers and quotations from the English translation: *General Theory of Knowledge.* Albert E. Blumberg, trans. La Salle, IL: Open Court, 1985.

Schlick, Moritz. 1931. "Die Kausalität in der gegenwärtigen Physik." *Die Naturwissenschaften* 19, 145–62.

Schlick, Moritz. 1979. *Philosophical Papers,* 2 vols. edited by Henk L. Mulder and Barbara F. B. van de Velde-Schlick, translated by Peter Heath. Vienna Circle Collection, vol. 11. Dordrecht and Boston: D. Reidel.

Schneider, Ilse. 1921. *Das Raum-Zeit Problem bei Kant und Einstein.* Berlin: Springer.

Sellien, Ewald. 1919. *Die erkenntnistheoretische Bedeutung der Relativitätstheorie. Kant-Studien* Ergänzungshefte, no. 48. Berlin: Reuther & Reichard.

Solovine, Maurice, ed. 1956. *Albert Einstein. Lettres à Maurice Solovine.* Paris: Gauthier-Villars. Page numbers and quotations from the English trans.: *Albert Einstein: Letters to Solovine.* Wade Baskin, trans. New York: Philosophical Library, 1987.

Speziali, Pierre, ed. 1972. *Albert Einstein-Michele Besso. Correspondence 1903–1955.* Paris: Hermann.

Stachel, John. 1980. "Einstein's Search for General Covariance, 1912–1915." Paper delivered at the Ninth International Conference on General Relativity, Jena, German Democratic Republic, 17 July 1980. First published in Howard and Stachel 1989, 63–100.

Stadler, August. 1874. *Kants Teleologie und ihre erkenntnistheoretische Bedeutung. Eine Untersuchung.* Berlin: Ferdinand Dümmler, Harrwitz & Gossmann.

Stadler, August. 1883. *Kants Theorie der Materie.* Leipzig: S. Hirzel.

Stadler, Friedrich. 1997. *Der Wiener Kreis. Ursprung, Entwicklung und Wirkung des Logischen Empirismus im Kontext.* Frankfurt am Main: Suhrkamp. English trans.: *The Vienna Circle: Studies in the Origins, Development, and Influence of Logical Empiricism*. Camilla Nielson, trans. Vienna: Springer, 2001.

Stuart, Henry. 1910. "Dühring, Eugen Karl." In *Encyclopædia Britannica*, 11th ed. (Vol. 8, 649). New York: Encyclopædia Britannica.

Talmey, Max. 1932. *The Relativity Theory Simplified and the Formative Period of Its Inventor*. New York: Falcon Press.

Veblen, Oswald. 1904. "A System of Axioms for Geometry." *American Mathematical Society. Transactions* 5, 343–84.

von Kloeden, Wolfdietrich. 2003. "Riehl, Alois." In *Neue Deutsche Biographie*, vol 21. Berlin: Duncker & Humblot, 586–87.

Westfall, Richard S. 1980. *Never at Rest: A Biography of Isaac Newton*. New York: Cambridge University Press.

The Contributors

Barry Clark Barish is an American experimental physicist and Nobel Laureate. He is Linde Professor of Physics, Emeritus, at California Institute of Technology. A leading expert on gravitational waves, Barish was awarded the Nobel Prize in Physics in 2017 along with Rainer Weiss and Kip Thorne "for decisive contributions to the LIGO detector and the observation of gravitational waves." In 2018 he joined the faculty at University of California, Riverside, becoming the university's second Nobel Prize winner on the faculty.

Alessandra Buonanno is the director of the Astrophysical and Cosmological Relativity Department at the Max Planck Institute for Gravitational Physics, and holds a College Park professorship at the University of Maryland. She is a leading theorist in the field of gravitational-wave physics, and a Principal Investigator of the LIGO Scientific Collaboration. Her work at the intersection of analytical-relativity modeling and numerical-relativity simulations was essential in the detection of gravitational waves from binary systems and the physical interpretation of the signals. For her contributions to LIGO and Virgo discoveries, she was awarded the 2018 Leibniz Prize—the most prestigious research prize in Germany awarded by the German National Science Foundation. She is a Fellow of the International Society on General Relativity and Gravitation, and of the American Physical Society.

Diana Kormos Buchwald, initially trained as a physical chemist, studied the history of science at Harvard and joined the faculty of the California Institute of Technology in 1989, where she is the Robert M. Abbey Professor of History. Since her appointment as General Editor and Director of the Einstein Papers Project in 2000, the editorial team has published eight volumes of *The Collected Papers of Albert Einstein* in documentary and in English language editions with Princeton University Press.

Jed Buchwald (editor) is Doris and Henry Dreyfuss Professor of History at Caltech and Director of the Caltech-Huntington Advanced Research Institute in the History of Science and Technology. He edits or co-edits four book series and two journals in the history of science and has edited or co-edited eight books. Buchwald has written three books published by the University of Chicago Press on the history of electrodynamics and optics in the nineteenth century. He is the co-author of three others: *The Zodiac of Paris*, on a controversy concerning the age of Egypt in the early nineteenth century; *Newton and the Origin of Civilization*, on Newton's use of astronomy to re-date the past; and *The Riddle of the Rosetta*, on the decipherment of Egyptian hieroglyphs (all Princeton).

Harry Collins is Distinguished Research Professor at Cardiff University. He is an elected Fellow of the British Academy and winner of the Bernal Prize for social studies of science. His c25 books cover, among other things, sociology of scientific knowledge, artificial intelligence, the nature of expertise, tacit knowledge, and technology in sport. His contemporaneous study of the detection of gravitational waves has been continuing since 1972, and he has written four books and many papers on the topic. He is currently looking at the impact of the coronavirus lockdown on science due to the ending of face-to-face conferences and workshops.

Don Howard is a professor of philosophy at the University of Notre Dame, where he is also a past director of the Notre Dame Graduate Program in History and Philosophy of Science and Notre Dame's John J. Reilly Center for Science, Technology, and Values. A former assistant editor on *The Collected Papers of Albert Einstein*, Howard works mainly on the history and philosophy of modern physics.

Daniel Kennefick received his PhD in Physics from Caltech in 1997. Since 2003 he has been in the physics department at the University of Arkansas. In 2007 he published a book on the history of gravitational waves titled *Traveling at the Speed of Thought*; he is also the author of the recently published *No Shadow of a Doubt*, about the 1919 eclipse expeditions that tested general relativity, and a co-author of *An Einstein Encyclopedia* (all Princeton). His research concentrates on three areas—gravitational waves, galactic structure, and history of physics.

Jürgen Renn is Director at the Max Planck Institute for the History of Science in Berlin (MPIWG). His research focuses on long-term development of knowledge while taking into account processes of globalization and the historical origins and co-evolutionary dynamics leading into the Anthropocene. Over more than two decades at the MPIWG, his research projects have dealt with a number of different historical subjects, including the development of mechanics from antiquity until the twentieth century and the origin and development of the general theory of relativity and of quantum theory. His most recent publication is *The Evolution of Knowledge: Rethinking Science for the Anthropocene* (Princeton).

Tilman Sauer is professor of history of mathematics and the exact sciences at Johannes Gutenberg University in Mainz, Germany. After obtaining a Ph.D. in theoretical physics from the Free University in Berlin in 1994, he spent several years at the Max Planck Institute for Human Development and Education and at the Max Planck Institute for the History of Science in Berlin. He also held postdoctoral and teaching positions at the Universities of Göttingen and Bern. From 2001 to 2013 he was a member of the editorial team of the Collected Papers of Albert Einstein and senior research faculty in history at Caltech.

Kip Thorne is the Feynman Professor of Theoretical Physics, Emeritus, at the California Institute of Technology. From 1967 to 2009 he led a research group at Caltech working on relativity, astrophysics, and gravitational-wave science and technology, and mentored 53 PhD students and about 60 postdoctoral students. With Rainer Weiss and Ronald Drever he co-founded the LIGO Project. For his contributions to LIGO he was awarded the 2017 Nobel Prize in Physics, together with MIT's Weiss and Caltech's Barish. Since 2009 his primary focus has been the interface between science and the arts, including the Hollywood movie *Interstellar*, his accompanying book *The Science of Interstellar*, and a forthcoming book of his verse and Lia Halloran's paintings titled *The Warped Side of our Universe*.

Index